THE ENCYCLOPEDIA OF PSYCHOACTIVE DRUGS

SERIES 1

The Addictive Personality
Alcohol and Alcoholism
Alcohol: *Customs and Rituals*
Alcohol: *Teenage Drinking*
Amphetamines: *Danger in the Fast Lane*
Barbiturates: *Sleeping Potions or Intoxicants?*
Caffeine: *The Most Popular Stimulant*
Cocaine: *A New Epidemic*
Escape from Anxiety and Stress
Flowering Plants: *Magic in Bloom*
Getting Help: *Treatments for Drug Abuse*
Heroin: *The Street Narcotic*
Inhalants: *The Toxic Fumes*

LSD: *Visions or Nightmares?*
Marijuana: *Its Effects on Mind & Body*
Methadone: *Treatment for Addiction*
Mushrooms: *Psychedelic Fungi*
Nicotine: *An Old-Fashioned Addiction*
Over-The-Counter Drugs: *Harmless or Hazardous?*
PCP: *The Dangerous Angel*
Prescription Narcotics: *The Addictive Painkillers*
Quaaludes: *The Quest for Oblivion*
Teenage Depression and Drugs
Treating Mental Illness
Valium: *and Other Tranquilizers*

SERIES 2

Bad Trips
Brain Function
Case Histories
Celebrity Drug Use
Designer Drugs
The Downside of Drugs
Drinking, Driving, and Drugs
Drugs and Civilization
Drugs and Crime
Drugs and Diet
Drugs and Disease
Drugs and Emotion
Drugs and Pain
Drugs and Perception
Drugs and Pregnancy
Drugs and Sexual Behavior

Drugs and Sleep
Drugs and Sports
Drugs and the Arts
Drugs and the Brain
Drugs and the Family
Drugs and the Law
Drugs and Women
Drugs of the Future
Drugs Through the Ages
Drug Use Around the World
Legalization: *A Debate*
Mental Disturbances
Nutrition and the Brain
The Origins and Sources of Drugs
Substance Abuse: *Prevention and Treatment*
Who Uses Drugs?

NICOTINE

EDITOR, WRITER
OF UPDATED MATERIAL

Ann Keene

GENERAL EDITOR
OF UPDATING PROJECT

Professor Paul R. Sanberg, Ph.D.

Department of Psychiatry, Neurosurgery,
Physiology, and Biophysics
University of Cincinnati College of Medicine; and
Director of Neuroscience, Cellular Transplants, Inc.

GENERAL EDITOR

Professor Solomon H. Snyder, M.D.

Distinguished Service Professor of
Neuroscience, Pharmacology, and Psychiatry at
The Johns Hopkins University School of Medicine

ASSOCIATE EDITOR

Professor Barry L. Jacobs, Ph.D.

Program in Neuroscience, Department of Psychology,
Princeton University

SENIOR EDITORIAL CONSULTANT

Jerome H. Jaffe, M.D.

Director of The Addiction Research Center,
National Institute on Drug Abuse

THE ENCYCLOPEDIA OF PSYCHOACTIVE DRUGS

NICOTINE

An Old-Fashioned Addiction

JACK E. HENNINGFIELD, Ph.D.

The Johns Hopkins University School of Medicine

and

The Addiction Research Center
National Institute on Drug Abuse

CHELSEA HOUSE PUBLISHERS

NEW YORK   PHILADELPHIA

The author gratefully acknowledges the efforts of Judy Murphy and Robert Hutchings at the U.S. Government Office on Smoking and Health for providing detailed information and literature on the causes and effects of cigarette smoking.

Chelsea House Publishers

EDITOR-IN-CHIEF: Remmel Nunn
MANAGING EDITOR: Karyn Gullen Browne
PICTURE EDITOR: Adrian G. Allen
ART DIRECTOR: Maria Epes
MANUFACTURING MANAGER: Gerald Levine
SYSTEMS MANAGER: Lindsey Ottman
PRODUCTION MANAGER: Joseph Romano

THE ENCYCLOPEDIA OF PSYCHOACTIVE DRUGS
EDITOR OF UPDATED MATERIAL: Ann Keene

STAFF FOR NICOTINE: AN OLD-FASHIONED ADDICTION
PRODUCTION EDITOR: Marie Claire Cebrián
LAYOUT: Bernard Schleifer
APPENDIXES AND TABLES: Gary Tong
PICTURE RESEARCH: Susan Quist, Jonathan Shapiro

UPDATED 1992
3 5 7 9 8 6 4

Library of Congress Cataloging in Publication Data
Henningfield, Jack E.
    Nicotine: an old-fashioned addiction.
    (The Encyclopedia of psychoactive drugs)
    Bibliography: p.
    Includes index.
    Summary: Presents the latest information on the physical and psychological effects of smoking, the reasons people smoke, and the treatment for quitting this drug addiction.
    1. Smoking—Juvenile literature.   2. Tobacco habit—Juvenile literature.
3. Nicotine—Juvenile literature. [1. Smoking.   2. Nicotine] I. Title.
II. Series.
RC567.H46   1984   613.8'5   84-14956
ISBN 0-87754-751-3
     0-7910-0768-5 (pbk.)

CONTENTS

An American Cancer Society poster cleverly emphasizes some of the negative aspects of smoking.

FOREWORD

Since the 1960s, the abuse of psychoactive substances—drugs that alter mood and behavior—has grown alarmingly. Many experts in the fields of medicine, public health, law enforcement, and crime prevention are calling the situation an epidemic. Some legal psychoactive substances—alcohol, caffeine, and nicotine, for example—have been in use since colonial times; illegal ones such as heroin and marijuana have been used to a varying extent by certain segments of the population for decades. But only in the late 20th century has there been widespread reliance on such a variety of mind-altering substances—by youth as well as by adults.

Day after day, newspapers, magazines, and television and radio programs bring us the grim consequences of this dependence. Addiction threatens not only personal health but the stability of our communities and currently costs society an estimated $180 billion annually in the United States alone. Drug-related violent crime and death are increasingly becoming a way of life in many of our cities, towns, and rural areas alike.

Why do people use drugs of any kind? There is one simple answer: to "feel better," physically or mentally. The antibiotics your doctor prescribes for an ear infection destroy the bacteria and make the pain go away. Aspirin can make us more comfortable by reducing fever, banishing a headache, or relieving joint pain from arthritis. Cigarettes put smokers at ease in social situations; a beer or a cocktail helps a worker relax after a hard day on the job. Caffeine, the most widely used drug in America, wakes us up in the morning and overcomes fatigue when we have exams to study for or a long drive to make. Prescription drugs, over-the-counter remedies, tobacco products, alcoholic beverages, caffeine products—all of these are legally available substances that have the capacity to change the way we feel.

But the drugs causing the most concern today are not found in a package of NoDoz or in an aspirin bottle. The drugs that government and private agencies are spending billions of dollars to overcome in the name of crime prevention, law enforcement, rehabilitation, and education have names like crack, angel dust, pot, horse, and speed. Cocaine, PCP, marijuana, heroin, and amphetamines can be very dangerous indeed, to both users and those with whom they live, go to school, and work. But other mood- and mind-altering substances are having a devastating impact, too—especially on youth.

Consider alcohol: The minimum legal drinking age in all 50 states is now 21, but adolescent consumption remains high, even as a decline in other forms of drug use is reported. A recent survey of high school seniors reveals that on any given weekend one in three seniors will be drunk; more than half of all high school seniors report that they have driven while they were drunk. The average age at which a child has his or her first drink is now 12, and more than 1 in 3 eighth-graders report having been drunk at least once.

Or consider nicotine, the psychoactive and addictive ingredient of tobacco: While smoking has declined in the population as a whole, the number of adolescent girls who smoke has been steadily increasing. Because certain health hazards of smoking have been conclusively demonstrated—its relationship to heart disease, lung cancer, and respiratory disease; its link to premature birth and low birth weight of babies whose mothers smoked during pregnancy—the long-term effects of such a trend are a cause for concern.

Studies have shown that almost all drug abuse begins in the preteen and teenage years. It is not difficult to understand why: Adolescence is a time of tremendous change and turmoil, when teenagers face the tasks of discovering their identity, clarifying their sexual roles, asserting their independence as they learn to cope with authority, and searching for goals. The pressures—from friends, parents, teachers, coaches, and one's own self—are great, and the temptation to want to "feel better" by taking drugs is powerful.

Psychoactive drugs are everywhere in our society, and their use and misuse show no sign of waning. The lack of success in the so-called war on drugs, begun in earnest in the 1980s, has shown us that we cannot "drug proof" our homes, schools, workplaces, and communities. What we can do, however, is make available the latest information on these substances and their effects and ask that those reading it consider the information carefully.

The newly updated ENCYCLOPEDIA OF PSYCHOACTIVE DRUGS, specifically written for young people, provides up-to-date information on a variety of substances that are widely abused in today's society. Each volume is devoted to a specific substance or pattern of abuse and is designed to answer the questions that young readers are likely to ask about drugs. An individualized glossary in each volume defines key words and terms, and newly enlarged and updated appendixes include recent statistical data as well as a special section on AIDS and its relation to drug abuse. The editors of the ENCYCLOPEDIA OF PSYCHOACTIVE DRUGS hope this series will help today's adolescents make intelligent choices as they prepare for maturity in the 21st century.

Ann Keene, Editor

An American Cancer Society poster asks a question designed to prevent young nonsmokers from ever starting and to encourage young smokers to quit before they become heavily addicted.

INTRODUCTION

USES AND ABUSES

Jack H. Mendelson, M.D.
Nancy K. Mello, Ph.D.
Alcohol and Drug Abuse Research Center
Harvard Medical School—McLean Hospital

Human beings are endowed with the gift of wizardry, a talent for discovery and invention. The discovery and invention of substances that change the way we feel and behave are among our special accomplishments, and like so many other products of our wizardry, these substances have the capacity to harm as well as to help.

Consider alcohol—available to all and recognized as both harmful and pleasure inducing since biblical times. The use of alcoholic beverages dates back to our earliest ancestors. Alcohol use and misuse became associated with the worship of gods and demons. One of the most powerful Greek gods was Dionysus, lord of fruitfulness and god of wine. The Romans adopted Dionysus but changed his name to Bacchus. Festivals and holidays associated with Bacchus celebrated the harvest and the origins of life. Time has blurred the images of the Bacchanalian festival, but the theme of drunkenness as a major part of celebration has survived the pagan gods and remains a familiar part of modern society. The term *Bacchanalian festival* conveys a more appealing image than "drunken orgy" or "pot party," but whatever the label, some of the celebrants will inevitably start up the "high" escalator to the next plateau. Once there, the de-escalation is often difficult.

According to reliable estimates, 1 out of every 10 Americans develops a serious alcohol-related problem sometime in his or her lifetime. In addition, automobile accidents caused by drunken drivers claim the lives of more than 20,000 people each year, and injure 25 times that number. Many of the victims are gifted young people just starting out in adult life. Hospital emergency rooms abound with patients seeking help for alcohol-related injuries.

Who is to blame? Can we blame the many manufacturers who produce such an amazing variety of alcoholic beverages? Should we blame the educators who fail to explain the perils of intoxication or so exaggerate the dangers of drinking that no one could possibly believe them? Are friends to blame— those peers who urge others to "drink more and faster," or the macho types who stress the importance of being able to "hold your liquor?" Casting blame, however, is hardly constructive, and pointing the finger is a fruitless way to deal with problems. Alcoholism and drug abuse have few culprits but many victims. Accountability begins with each of us, every time we choose to use or to misuse an intoxicating substance.

It is ironic that some of our earliest medicines, derived from natural plant products, are used today to poison and to intoxicate. Relief from pain and suffering is one of society's many continuing goals. More than 3,000 years ago, the Therapeutic Papyrus of Thebes, one of our earliest written records, gave instructions for the use of opium in the treatment of pain. Opium, in the form of its major derivative, morphine, remains one of the most powerful drugs we have for pain relief. But opium, morphine, and similar compounds, such as heroin, have also been used by many to induce changes in mood and feeling. Another example of a natural substance that has been misused is the coca leaf, which for centuries was used by the Indians of Peru to reduce fatigue and hunger. Its modern derivative, cocaine, has important medical use as a local anesthetic. Unfortunately, its increasing abuse in recent years has reached epidemic proportions.

The purpose of this series is to provide information about the nature and behavioral effects of alcohol and drugs and the probable consequences of their use. The authors believe that up-to-date, objective information about alcohol and drugs will help readers make better decisions about the wisdom of their use. The information presented here (and in other books in this series) is based on many clinical and laboratory studies and observations by people from diverse walks of life.

Over the centuries, novelists, poets, and dramatists have provided us with many insights into the effects of alcohol and drug use. Physicians, lawyers, biologists, psychologists, and social scientists have contributed to a better understanding of the causes and consequences of using these substances. The authors in this series have attempted to gather and condense all the latest information about drug use. They have also described the sometimes wide gaps in our knowledge and have suggested some new ways to answer many difficult questions.

How, for example, do alcohol and drug problems get started? And what is the best way to treat them when they do? Not too many years ago, alcoholics and drug abusers were regarded as evil, immoral, or both. Many now believe that these persons suffer from very complicated diseases involving deep psychological and social problems. To understand how the disease begins and progresses, it is necessary to understand the nature of the substance, the behavior of the afflicted person, and the characteristics of the society or culture in which that person lives.

The diagram below shows the interaction of these three factors. The arrows indicate that the substance not only affects the user personally but the society as well. Society influences attitudes toward the substance, which in turn affect its availability. The substance's impact upon the society may support or discourage the use and abuse of that substance.

SUBSTANCE
(ALCOHOL OR DRUG)

PERSON ◄───────────► SOCIETY

Although many of the social environments we live in are very similar, some of the most subtle differences can strongly influence our thinking and behavior. Where we live, go to school and work, whom we discuss things with—all influence our opinions about drug use. Yet we also share certain commonly accepted beliefs that outweigh any differences in our attitudes. The authors in this series have tried to identify and discuss the central, most crucial issues concerning drug use.

Regrettably, human wizardry in developing new substances in medical therapeutics has not always been paralleled by intelligent usage. Although we do know a great deal about the effects of alcohol and drugs, we have yet to learn how to impart that knowledge, especially to young adults.

Does it matter? What harm does it do to smoke a little pot or have a few beers? What is it like to be intoxicated? How long does it last? Will it make me feel really fine? Will it make me sick? What are the risks? These are but a few of the questions answered in this series, which we hope will enable the reader to make wise decisions concerning the crucial issue of drugs.

Information sensibly acted upon can go a long way toward helping everyone develop his or her best self. As one keen and sensitive observer, Dr. Lewis Thomas, has said,

> *There is nothing at all absurd about the human condition. We matter. It seems to me a good guess, hazarded by a good many people who have thought about it, that we may be engaged in the formation of something like a mind for the life of this planet. If this is so, we are still at the most primitive stage, still fumbling with language and thinking, but infinitely capacitated for the future. Looked at this way, it is remarkable that we've come as far as we have in so short a period, really no time at all as geologists measure time. We are the newest, the youngest, and the brightest thing around.*

Every year cigarettes kill more Americans than were killed in World War I, the Korean War, and Vietnam combined; nearly as many as died in battle in World War II. Each year cigarettes kill five times more Americans than do traffic accidents. Lung cancer alone kills as many as die on the road. The cigarette industry is peddling a deadly weapon. It is dealing in people's lives for financial gain.

—from an address by United States Senator
Robert F. Kennedy of New York
at the First World Conference on Smoking and Health,
New York, NY, September 11, 1967

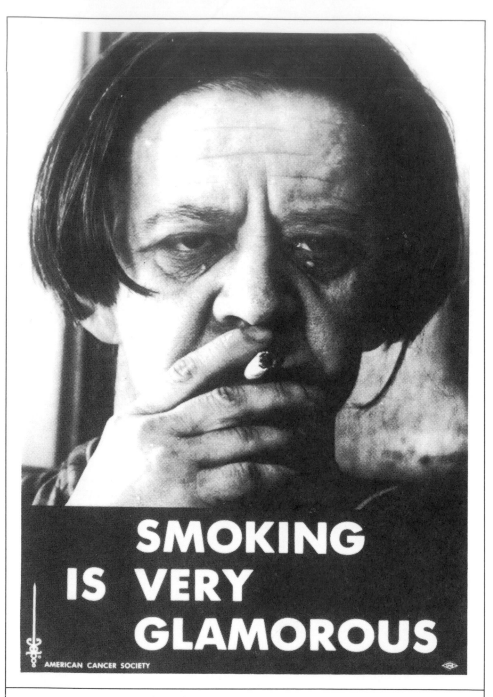

SMOKING
IS VERY
GLAMOROUS

AMERICAN CANCER SOCIETY

An American Cancer Society poster. The public health lobby often hires advertising agencies to communicate the antismoking message, taking advantage of a medium also exploited fully by the tobacco industry.

Almost one-third of adult Americans smoke cigarettes. This is a remarkable statistic considering that most people believe smoking is harmful to their health, and most smokers say they would quit if they knew of an effective way.

Why do so many people smoke cigarettes? Is smoking a "voluntary pleasure" as claimed by the tobacco industry? A behavioral habit akin to eating potato chips? A form of self-medication for nervousness and weight control? Or a form of drug abuse? There are elements of truth in all of these explanations. Some, however, are more valid than others.

There is now considerable evidence about the harmful effects of smoking. It is clear that if people smoked less, tobacco-related diseases and deaths would decrease. So it is very important to understand why people smoke.

The study of the medical effects of smoking began in earnest in the 1950s, and the First Surgeon General's Report on Smoking and Health was released in 1964. But the study of the *causes* of smoking didn't begin until much later. Most of it has taken place in the last 15 years.

There are several reasons why the study of the causes of smoking lagged so far behind the study of its effects. For one thing researchers couldn't justify spending time and money on the causes until it was shown that the effects were harmful.

Also, the reasons for smoking are not easy to pinpoint. One of the major questions is whether there are physical and psychological factors that cause or influence people to smoke—factors they aren't even aware of. So you can't just ask smokers why they smoke. Their answers may be true as far as they go, but they are probably only part of the story. To study a complicated behavior like smoking you need sophisticated research techniques, and such techniques have not been available until recently.

Research into causes was also delayed because of an earlier theory about smoking. This theory held that smoking was a voluntary behavior and that if people knew it was harmful, they would quit. Government and health organizations made a serious effort to convince people that smoking

A root-like malignant cancer cell seen through an electron microscope. Smoking is a contributing factor in 30% of all deaths from cancer in the United States.

was harmful. They showed movies of lung operations and publicized statistics on deaths and diseases caused by tobacco. Some people cut down or quit, but most kept right on smoking. This provided a clue to the nature of smoking. Continuing a behavior that is known to be damaging is a hallmark of drug abuse.

This book presents the most up-to-date thinking and scientific knowledge about the physical and psychological effects of smoking, the reasons people smoke, and treatments for smoking. The current view of cigarette smoking is that a better understanding of smoking behavior will lead to better forms of treatment for those who want to quit.

Aversion therapy exaggerates the more unpleasant aspects of an activity normally considered pleasurable. Here smoke is blown into the face of a smoker who wants to quit.

Tobacco was discovered in the New World, and became a major commodity during the 16th century. European traders advertised their product with wooden effigies that became known as "cigar store Indians." The statues suggested the exotic and faraway source of tobacco, which by 1600 had been introduced to Sweden, Russia, Turkey, Egypt, Persia, India, and China. So popular was the leaf in the New World that it was legal tender for the payment of wages, debts, and taxes in many colonies.

CHAPTER 1

TOBACCO'S PAST

*T*he Huron Indians of North America have passed down a legend concerning the origin of tobacco. According to the legend, there was once a great famine, when all the lands were barren. The Great Spirit sent a naked girl to restore the land and save his people. Where she touched the ground with her right hand, potatoes grew and the earth was fertile. Where she touched the ground with her left hand, corn sprang forth, bringing green to the lands and filling all stomachs. Finally, the naked messenger of the Great Spirit sat down; and from her place of rest grew tobacco.

There are two interpretations of this tale. One is that tobacco was a gift like corn and potatoes and was meant to provide food for the mind of man. The second is that since tobacco was given by the seat of the messenger, it was intended as a message (or possibly a curse) that the gifts of food were not without their price.

Whichever interpretation is correct, the use of tobacco was firmly established among North American Indians by the time Columbus arrived. The early explorers were amazed to discover the Indians putting little rolls of dried leaves into their mouths and then setting them afire. Some Indians carried pipes in which they burned the same leaves so they could "drink" the smoke. It was also apparent that tobacco was essential to many religious and social rituals and that it was a habit not given up easily.

Tobacco Takes Hold

In the 16th century two British sea captains persuaded three Indians to return with them to London. The Indians brought substantial supplies of tobacco to sustain them through their voyage and stay. That trip may have marked the birth of the Indians' revenge against the white invaders of their land, for along the way some crew members tried inhaling the tobacco smoke. Many enjoyed it and soon found it hard to stop. To supply their own needs for tobacco, the explorers of the 16th and 17th centuries kept fields around the Horn of Africa, in Europe, and in the Americas. Magellan's crew smoked tobacco and left seeds in the Philippines and other ports of call. The Dutch brought tobacco to the Hottentots. The Portuguese brought it to the Polynesians. Soon, wherever sailors went—in Asia, Africa, even Australia—tobacco was waiting.

By the beginning of the 17th century, small plots of tobacco cultivation had become plantations throughout the world. Wherever it was grown, of course, the inhabitants also tried smoking it, thus expanding its use even further. Smoking spread almost like a contagious disease, from a few

A Mayan Indian smokes a string-tied cigar.

individuals to entire populations.

Early tobacco users quickly learned what the Indians had long known: once you started, you couldn't stop without severe discomfort and a powerful urge to resume the habit. Furthermore, these needs could only be satisfied by tobacco, which had to be used in certain ways. It did not have to be smoked. It could be chewed or ground to a powder and inhaled as "snuff." Simply eating the raw plant, however, did not provide relief or pleasure, and no other substance seemed to be an adequate substitute.

Opposition to the Use of Tobacco

As tobacco was introduced to one empire after another, a similar pattern of response occurred. What happened in England, under King James I, at the beginning of the 17th century, was typical. King James greatly opposed the use of tobacco. He decried its use as unhealthy and immoral,

Sir Walter Raleigh's servant discovers too late that he is extinguishing not a fire but Sir Walter's pipe. Raleigh, a British explorer and adventurer, made tobacco fashionable in England during the late 1500s.

and he urged its banishment. However, even among the royal court tobacco had its dedicated followers: men like Sir Walter Raleigh made tobacco use both fashionable and a mark of distinction.

Attempts to restrict supplies only increased the value of tobacco and soon it was worth its weight in silver. Finally, in one of the earliest recorded attempts at prohibition by economics, King James increased the tobacco duty tax by 4,000%. The only real consequence was to help stimulate a flourishing smuggling trade.

In the end, the economic issues conquered the rulers and not the plant. Seeing that the people would pay nearly any price for tobacco, monopolies were started so that the government could benefit from the desires of its people.

A South American Indian medicine man summons up the magical powers of tobacco to drive out the spirit which causes disease. Many primitive societies considered tobacco therapeutic.

Taxation policies were more carefully implemented and the government itself was soon dependent on the trading of tobacco.

This general pattern of disapproval, failed attempts at prohibition, and economic gain by taxation was repeated in Italy, France, Russia, Prussia, and then the United States. As governments became convinced of the dangers of tobacco use, taxes were raised, providing the dual benefit that the conscience of the government was cleansed while its income was enhanced as people continued to smoke. In the United States today smokers spend about $30 billion per year; a pack-a-day smoker spends more than $700 per year. More than one-third of that is for state and federal taxes (see Figure 1).

A COVNTERBLASTE
TO TOBACCO.

HAt the manifold abufes of this vile cuftome of *Tobacco* taking, may the better be efpied, it is fit, that firft you enter into confideration both of the firft originall thereof, and likewife of the reafons of the firft entry thereof into this Countrey. For certainely as fuch cuftomes, that haue their firft inftitution either from a godly, neceffary, or honourable ground, and are firft brought in, by the meanes of fome worthy, vertuous, and great Perfonage, are euer, and moft iuftly, holden in great and reuerent eftimation and account, by all wife, vertuous, and temperate fpirits : So fhould it by the contrary, iuftly bring a great difgrace into that fort of cu-

The opening page of an antismoking tract by King James I of England in the early 17th century.

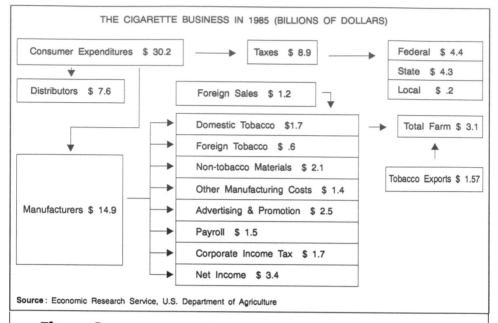

Figure 1. *The chart shows how the $30 billion tobacco industry was distributed in the United States in 1985. Today more than 50 brands of cigarettes are offered for sale in the United States, packaged in more than 200 varieties and sizes.*

As a "revenge of the Indians" the spread of tobacco makes the introduction of syphilis into the New World pale by comparison when the enormous toll in death and disease is considered. Tobacco addiction has shared with alcoholism the fact that it was a consequence of a behavior never to be eliminated, either by law, taxation, or papal mandate. Perhaps the most dramatic description of the Indians' revenge was given by "the tobacco fiend" at the court of Lucifer in an 18th century epic:

> *Thus do I take revenge in full upon the Spaniards for all their cruelty to the Indians; since by acquainting their conquerors with the use of tobacco I have done them greater injury than even the King of Spain through his agents ever did his victims; for it is both more honorable and more natural to die by a pike thrust or a cannon ball than from the ignoble effects of poisonous tobacco.*

Tobacco was widely used in religious rituals by the Aztec Indians of central Mexico (who were conquered by the explorer Hernando Cortés in 1519). In this illustration, an idol dressed as a god holds a cigarette.

Tobacco crops in West Virginia await harvesting. Between 1968 and 1981 America's cigarette market grew from $10.1 billion to $18 billion. Although the number of smokers declined during that period, those who did smoke were smoking more than ever before.

CHAPTER 2

WHAT IS IN TOBACCO SMOKE?

What is the nature of this substance which has caused so much controversy and is of such concern to governments, religious bodies, and science? The tobacco plant is a member of the vegetable family *Solanaceae*. The plant was named *Nicotiana tabacum* in honor of the French ambassador to Portugal in the 1580s, Jean Nicot, who believed the plant had medicinal value and encouraged its cultivation. One chemical constituent of tobacco is nicotine. When cultivated with various chemical fertilizers and insecticides, processed into cigarettes, and finally burned, many other physical constituents result.

Primary Tobacco Smoke Constituents

Tobacco smoke contains thousands of elements. Most are delivered in such minute amounts that they are not usually considered in discussions of the medical effects of cigarette smoking. In fact, there are so many that it will take years of research to discover which constituents are harmful. Three of undisputed importance, however, are tar, carbon monoxide, and nicotine.

Tar is defined, rather arbitrarily, as the total particulate matter (TPM), minus water and nicotine, which is trapped by the Cambridge filter used in smoke collection machines. Persons who have used ventilated cigarette holders (e.g., One Step At A Time®) in an effort to quit smoking have probably noticed the accumulation in the filter of a thick, black material reminiscent of road tar. Tar, not present in unburned tobacco, is a product of organic matter being burned in the presence of air and water at a sufficiently high temperature. Tobacco products such as snuff and chewing tobacco do not deliver tar.

The Federal Trade Commission (FTC) figures for tar, which are sometimes printed on cigarette packages, do not reflect tar contained in the tobacco or even in the smoke. These estimates reflect the amount collected from the standard cigarette-smoking machines. The levels may be useful for cigarette comparisons, but are otherwise misleading to people who think that their intake of tar is mainly determined by their brand of cigarettes. One study showed that very low tar cigarettes with FTC ratings of a few milligrams delivered 15 to 20 milligrams when smoked the way a person might actually do so.

Tar is one of the major health hazards in cigarette smoking. It causes a variety of types of cancer in laboratory animals. Also, the minute separate particles fill the tiny air holes (the alveoli) in the lungs and contribute to respiratory problems such as emphysema. In light of these facts many cigarette manufacturers have reduced the tar yields in their cigarettes in an effort to provide "safer" cigarettes. Unfortunately, tar is important to the taste of the cigarette and the satisfaction derived from smoking. Thus, when many people smoke low tar cigarettes, to get maximum enjoyment they inhale so deeply that they defeat the purpose of this type of cigarette. It is ironic that cigarettes engineered to deliver low tar yields when smoked by machines deliver higher yields when smoked by people.

Carbon monoxide (CO) is a gas that results when materials are burned. Carbon monoxide production is increased by restricting the oxygen supply, as is the case inside a

Two thousand cigarettes, the number stacked on the table, produce the amount of tar contained in the flask. Tar is a major reason why cigarette smoking is dangerous.

cigarette. Carbon monoxide is also produced by internal combustion engines (automobile) and even by gas stoves and ovens. Like carbon dioxide (CO_2), which also results from burning, carbon monoxide easily passes from the alveoli of the lungs into the blood stream. There it combines with hemoglobin to form carboxyhemoglobin (COHb). Hemoglobin is that portion of the blood which normally carries carbon dioxide out of the body (CO_2 is produced by normal metabolic processes) and oxygen back into the body. When the hemoglobin is all bound up by either carbon monoxide or carbon dioxide, a shortage of oxygen may result.

A critical difference between carbon monoxide and carbon dioxide is that the former binds much more tightly to hemoglobin and is very slow to be removed. Thus the blood can accumulate rather high levels of carbon monoxide and slowly starve the body of oxygen. When the cardiac system detects insufficient levels of oxygen, the heart may begin to flutter and operate inefficiently. In extreme cases a heart attack may result.

Figure 2 shows carbon monoxide levels in cigarette smokers who smoked different numbers of cigarettes. These values were collected by having a group of smokers smoke one of their usual brand of cigarettes every 30 minutes until they had smoked either 5 or 10 cigarettes. The figure shows that each cigarette causes a brief boost in the CO level which lasts for a few minutes and then declines until the next cigarette is smoked.

However, each cigarette adds slightly to the overall

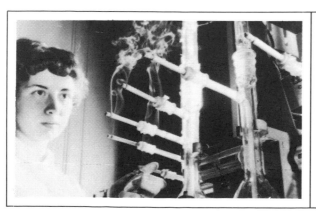

A laboratory technician analyzes the effectiveness of filters in capturing tar, nicotine, and other elements of cigarette smoke. While simulation of the many variables remains difficult, the use of computer models in recent years has substantially aided the research effort.

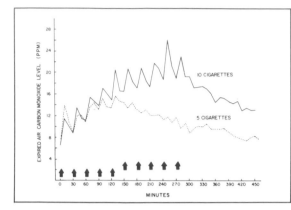

Figure 2. *Average carbon monoxide (CO) levels are shown for eight volunteers who smoked one cigarette every 30 minutes as indicated by the arrows. On one day they smoked five cigarettes (dotted line), and on another day they smoked ten cigarettes (solid line).*

level. When people smoke normally, their CO levels are lowest in the morning and level off at their highest values by midday. The typical one-pack-per-day smoker achieves levels averaging between 25 and 35 parts per million. However, even these "average" smokers may hit short-term levels of greater than 100 parts per million. Firefighters are now routinely checked with portable CO analyzing machines while combating fires. If their levels exceed 150 parts per million they may be relieved and given oxygen, since even these generally healthy people run a risk of heart attacks.

Nicotine is a drug that occurs naturally in the leaves of *Nicotiana tabacum.* It is generally thought of as a stimulant since it provokes many nerve cells in the brain and heightens arousal. However, its effects are so complex that no simple label is completely accurate. For instance, by stimulating certain nerves in the spinal cord (Renshaw cells), nicotine relaxes many of the muscles of the body and can even depress knee reflexes. Its effects also vary depending on how much is smoked. For example, nerve cells that are stimulated by the nicotine from a few cigarettes may be depressed by smoking more cigarettes.

Nicotine closely resembles one of the substances that occur naturally in the body (acetylcholine), and the body has an efficient system to break nicotine down (detoxification) and eliminate it in the urine (excretion). In fact, when a given dose of nicotine is ingested, for instance by smoking, about one-half is removed from the blood stream within 15 to 30 minutes. Figure 3 shows the pattern of nicotine accumulation and elimination when cigarettes are smoked. Note that these patterns are similar to those observed for carbon monoxide (see Figure 2).

Another feature of nicotine is that it is well absorbed through the mucosae, the very thin skin of the nose or mouth which is dense with capillaries. This is why chewing tobacco and snuffing are such effective ways to ingest nicotine. In the form of cigarette smoke, nicotine transfers directly from the alveoli of the lungs into the arterial bloodstream and rushes directly to the brain. It requires less than 10 seconds for inhaled nicotine to reach the brain, so that even though the quantities are small, the effects may be strong because the delivery system is so efficient. This is even quicker than giving nicotine intravenously, since the venous blood supply must first pass through the heart, then into the arterial stream of the lungs and, finally, to the brain. In addition, smoking delivers a much higher dose of nicotine than other intake methods do. A highly concentrated dose of nicotine is carried in the blood through the arteries on a direct path from the lungs to the brain.

Repeated exposure to nicotine, when it is smoked, results in very rapid tolerance or diminished effect. That is, during the day, as cigarettes are smoked, the smoker get less and less of a psychological and physical effect—even though toxins are building up in the body. In fact, much of this tolerance is lost overnight. As a result, cigarette smokers often report that their first cigarette of the day "tastes the best" or may even make them lightheaded if smoked too quickly. As the day wears on and more cigarettes are smoked, people often smoke more out of habit or to avoid discomfort than for pleasure.

Research to date on the hazards of cigarette smoking suggests that nicotine is not as hazardous as the tar and

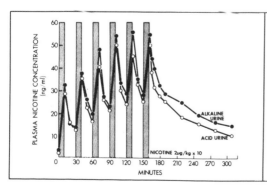

Figure 3. *Five volunteers were given injections of nicotine. Average nicotine levels in plasma are shown by the circles. Acidification of the urine resulted in more rapid nicotine excretion and lower plasma levels of nicotine, showing the body's capacity to eliminate nicotine levels in the bloodstream.*

Figure 4. *Automatic cigarette smoking machines continue to make a great contribution to smoking research. Technicians can modify various aspects of puffing, such as volume and duration of intake, at the push of a button. The smoke is drawn into the machine and stored in special containers.*

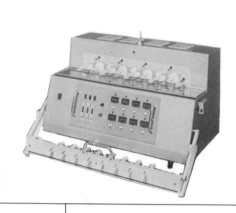

A beagle "smokes" a cigarette through a machine linked to its windpipe during an experiment investigating the connection between smoking and emphysema, a chronic lung disease.

A hospital employee in San Diego takes part in an experiment which sought to determine how quickly the lungs heal once a smoker has quit the habit. Recent research indicates that lung performance improves markedly soon after the termination of smoking.

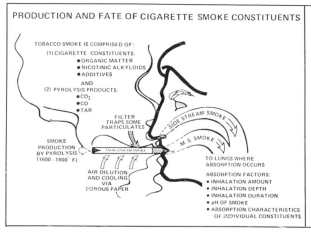

PRODUCTION AND FATE OF CIGARETTE SMOKE CONSTITUENTS

Figure 5. *The diagram illustrates what happens when a cigarette is smoked. As it burns, various substances are created, including carbon dioxide, carbon monoxide, and tar, that mingle with constituents that were already in the cigarette.*

carbon monoxide in cigarette smoke. Its role is more insidious, since people ingest these other substances (tar and CO) as a by-product of their efforts to obtain nicotine.

Other Constituents of Tobacco Smoke

Cigarette smoke is made up of both a gas phase and a particulate phase. Together they include more than 4,000 substances. Automatic cigarette-puffing machines have been devised to collect and to study smoke. Figure 4 shows such a machine, which might be used by research laboratories or cigarette manufacturers. The smoke is separated into the gas and solid (particulate) phases by passing it through a filter pad (Cambridge filter), which traps particles larger than one micrometer and collects the rest (gas phase) in a storage tank. The machines are calibrated to smoke the cigarettes the way a typical smoker might.

Figure 5 shows what happens during a puff. The unburned cigarette is made up of many organic (tobacco leaves, paper products, sugars, nicotine) and inorganic (water, radioactive elements, metals) materials. The tip of the burning cone in the center of the cigarette reaches a temperature of nearly 2,000 degrees Fahrenheit during a puff. This tiny blast furnace results in a miniature chemical plant, which uses the hundreds of available materials to produce many more. In fact, some of the most important parts of tobacco smoke (including tar and carbon monoxide) are not even present in an unburned cigarette, but rather are produced when a puff is taken.

Study of the smoke is made even more complicated since there are both sidestream and mainstream smoke which must be separately collected and studied. The mainstream smoke is collected from the stream of air passing through the center of the cigarette. It is filtered by the tobacco itself and perhaps further by a filter. It is also diluted by air passing through the paper (most modern cigarettes also have tiny ventilation holes which further dilute the smoke).

Sidestream smoke is that which escapes from the tip of the cigarette. It is not filtered by the cigarette and results from a slightly cooler burning process at the edge of the burning cone. Since the tobacco is therefore burned less completely, the sidestream smoke has more particulate (unburned material) in it.

Cigarette Engineering

The above process is complicated even further by the engineering efforts of the tobacco manufacturers. They specifically construct cigarettes in ways to control a wide range of factors: keeping the cigarette burning between puffs, reducing spoilage of the tobacco, altering the flavor of the smoke, and controlling the amounts of substances (tar and nicotine) measured by government agencies.

The porosity of cigarette paper is specifically controlled to regulate the amount of air that passes through and dilutes the smoke. Porosity also affects how rapidly the cigarette

Machinery representing the state of the art in cigarette manufacture at R.J. Reynolds Tobacco Company, in 1961. Increasingly sophisticated production methods gave America's tobacco industry 61% of the world's tobacco export market at that time.

burns. Phosphates are added to the paper to ensure steady and even burning.

Several kinds of additives are present in the tobacco itself. One type of additive is called a *humectant.* Humectants are chemicals that help retain the moisture (humidity) of the tobacco. This is important in how the tobacco burns. Humectants also affect the taste and temperature of the smoke. The most commonly used humectants are glycerol, D-sorbitol, and diethylene glycol. Humectants make up a small percentage of the total weight of the tobacco.

Another type of additive is called a *casing agent.* This helps blend the tobacco and hold it together. It also affects the flavor of the smoke and how quickly the tobacco burns. Most commonly used casing agents include sugars, syrups, licorice, and balsams. The amount of casing agents used ranges from about 5% of the total weight of the tobacco in cigarette tobacco to about 30% of the weight of pipe tobacco.

Specific *flavoring agents* are also added to the tobacco to control the characteristic taste of a cigarette. These include fruit extracts, menthol oils, spices, coca, aromatic materials, and synthetic additives. Flavor is also controlled by curing processes and, of course, the type of tobacco itself.

A variety of other substances are added at various stages of tobacco processing to retard spoilage and prevent tobacco larvae from developing into worms. In addition, metals such as nickel and potassium are taken up from the soil, as are pesticides and fertilizers used in tobacco farming. There are also radioactive elements such as potassium-40, lead-210, and radium-226, which result from fallout and the natural background.

Designed to attract young smokers, candy-flavored Nova *cigarettes went on the market in Japan in 1983. Manufacturers have continued to improve the naturally harsh taste of tobacco with many different flavoring additives.*

A detail from a temple carving shows a Mayan Indian priest smoking as part of a religious ceremony. Mayan sacred writings indicate that smoking was widespread in Central America long before Europeans arrived in the 16th century.

CHAPTER 3

TOBACCO AND HEALTH

*T*obacco has been used for its social and medicinal effects. In *Don Juan*, Sganarelle touts tobacco as "the passion of all proper people, and he who lives without tobacco has nothing else to live for. Not only does it refresh and cleanse men's brains, but it guides their souls in the ways of virtue, and by it one learns to be a man of honor."

Tobacco has been used at various times to treat headaches, asthma, gout, labor pains, and even cancer. Its diverse behavioral effects have also been noted. It was used by monks to suppress sexual drive. The philosopher Kant, on the other hand, regarded tobacco as a way of provoking sexual excitement. King James I wrote, "Being taken when they go to bed, it (tobacco) makes one sleep soundly, and yet being taken when a man is sleepy and drowsy, it will, as they say, awake his brain and quicken his understanding." (Generally an opponent of tobacco use, King James nonetheless made many accurate observations of the diverse effects of tobacco.)

Not all observers regarded tobacco so benignly. One of the strongest condemnations also came from King James I in his "Counterblaste to Tobacco." He called tobacco use a custom "lothesome to the eye, hatefull to the nose, harmfull to the brain, dangerous to the lungs, and in the black stinking fumes thereof, neerest resembling the horrible stinking

41

smoke of the pit that is bottomeless." It is interesting to point out that some have attributed King James' antismoking stance to his personal dislike of Sir Walter Raleigh, who strongly promoted the use of tobacco.

How Many Teenagers Smoke

Today a little less than one-third of adult Americans smoke cigarettes (see Table 1); most of them started smoking in their teens. In fact, of high school seniors who smoke regularly, less than 2% began in their senior year of high school and roughly two-thirds began by the ninth grade. While only about 12% of 12- to 17-year-olds now smoke regularly, most of them will continue to smoke and will learn to smoke stronger cigarettes at higher rates. (Young people smoke an average of about one-half pack per day compared to an average of about one pack per day for adults who are regular smokers.) Teenage smoking is also of concern since nicotine use, along with drinking alcohol, is a major precursor to illicit abuse of other psychoactive drugs. In 1985 a national survey

Table 1

Current Prevalence of Cigarette Use, 1974-1988 (Use in Past Month)							
				Percentages			
	1974	1976	1977	1979	1982	1985	1988
Teenagers (ages 12-17)	25.0	23.4	22.3	12.1	14.7	15.3	11.8
Young adults (ages 18-25	48.8	49.4	47.3	42.6	39.5	36.8	35.2
Older adults (ages 26 and over)	39.1	38.4	38.7	36.9	34.6	32.8	29.8

Source: National Institute on Drug Abuse, National Household Survey, 1988

revealed that among daily users of tobacco aged 12 to 17, about 40% had consumed intoxicating amounts of alcohol in the prior month and 27% had used marijuana more than 10 times in their lives. (Nonsmokers showed minimal use of both alcohol and marijuana.)

Some of the trends in smoking among young people have been surprising. In the 1960s young male smokers outnumbered young female smokers 2 to 1. By 1977 the numbers were about equal and had reached a level of more than 20% of teens. In the 1980s the prevalence of smoking among teenagers declined to its current level of about 12%.

Many teenagers perceive smoking to be much more prevalent than is actually the case, stating that "everyone smokes" or "it's the thing to do." In fact, about 1 in 9 persons aged 12 to 17 smokes cigarettes. The figures are even lower for youth involved in athletics or bound for college. The peak years for smoking among youth appear to have been in the early 1970s, when surveys of high school seniors reported that nearly 30% of them smoked daily. (See Table 2.) In 1989 that figure had dropped to about 19%.

These trends have been paralleled by changing attitudes toward smoking. Between 1975 and 1980 the percentage of high school seniors who viewed smoking as a "great risk" increased from 51% to almost 64%. Surveys during the 1980s indicate that more than 70% of high school seniors "disapprove" of smoking a pack or more of cigarettes per day, and more than 40% believe that smoking should be prohibited from certain public places. (See Table 3.)

This seven-year-old is only pretending to smoke during a tobacco auction in Georgia, but the desire to imitate adult behavior is one of the main reasons that many young people begin smoking.

It is also clear that the impact of the addictive nature of cigarettes has not been fully realized since less than 1% of high school seniors who smoke say that they will "definitely" be smoking in five years and only 13% say that they "probably" will be smoking in five years. The sad truth is that the majority will still be smoking five years later.

Health Consequences of Cigarette Smoking

Every day more than 2,000 American teenagers smoke their first cigarette; over the lifetimes of these 2,000, approximately 15 will be murdered, 20 will die in traffic accidents—and almost 500 will die from smoking-related diseases. Smoking is now responsible for more than one of every six deaths in the United States. In 1985, some 390,000 deaths were

Table 2

Cigarette Use Among High School Seniors, 1975-1989							
Class of 1975	Class of 1976	Class of 1977	Class of 1978	Class of 1979	Class of 1980	Class of 1981	Class of 1982
Percent reporting ever used:							
73.6	75.4	75.7	75.3	74.0	71.0	71.0	70.1
Percent reporting use in previous month:							
36.7	38.8	28.4	36.7	34.4	30.5	29.4	30.0
Percent reporting daily use in last 30 days:							
26.9	28.8	28.8	27.5	25.4	21.3	20.3	21.1
Percent reporting smoking half-pack or more per day:							
17.9	19.2	19.4	18.8	16.5	14.3	13.5	14.2

Source: National Institute on Drug Abuse, National High School Senior Survey, "Monitoring the Future," 1989

attributable to cigarette smoking. Of the five leading causes of death in the U.S.—heart disease, cancer, strokes, accidents, and chronic obstructive pulmonary diseases (COPD—mainly chronic bronchitis and emphysema)—cigarette smoking and other forms of tobacco use are implicated in all but one, accidents.

Heart Disease

Cigarette smoking is one of three major risk factors associated with heart disease, along with high blood pressure and high cholesterol levels. Heart disease is today the number-one killer in the United States; 765,000 people died of heart disease in the U.S. in 1986, and researchers estimate that 30 to 40 percent of these deaths are attributable to smoking.

Cigarette Use Among High School Seniors, 1975-1989

Class of 1983	Class of 1984	Class of 1985	Class of 1986	Class of 1987	Class of 1988	Class of 1989
Percent reporting ever used:						
70.6	69.7	68.8	67.6	67.2	66.4	65.7
Percent reporting use in previous month:						
30.3	29.3	30.1	29.6	29.4	28.7	28.6
Percent reporting daily use in last 30 days:						
21.2	18.7	19.5	18.7	18.7	18.1	18.9
Percent reporting smoking half-pack or more per day:						
13.8	12.3	12.5	11.4	11.4	10.6	11.2

Table 3

	Class of 1975	Class of 1976	Class of 1977	Class of 1978	Class of 1979
High School Seniors' Attitudes Toward Cigarette Smoking					
How much do you think people risk harming themselves (physically or in other ways) if they					
Smoke one or more packs of cigarettes per day? (Percentage saying "great risk")	51.3	56.4	58.4	59.0	63.0
Do you disapprove of people (who are 18 or older)					
Smoking one or more packs of cigarettes per day? (Percentage "disapproving")	67.5	65.9	66.4	67.0	70.3
Do you think that people (who are 18 or older) should be prohibited by law from					
Smoking cigarettes in certain specified public places? (Percentage saying "yes")	N.A.	N.A.	42.0	42.2	43.1
How do you think your close friends feel (or would feel) about your					
Smoking one or more packs of cigarettes per day? (Percentage saying friends disapprove)	63.6	N.A.	68.3	N.A.	73.4
How many of your friends would you estimate smoke cigarettes?					
Percentage saying none:	4.8	6.3	6.3	6.9	7.9
Percentage saying most or all:	41.5	36.7	33.9	32.2	28.6

N.A. = Not available
Source: National Institute on Drug Abuse, National High School Senior Survey, "Monitoring the Future"

High School Seniors' Attitudes Toward Cigarette Smoking

Class of 1980	Class of 1981	Class of 1982	Class of 1983	Class of 1984	Class of 1985	Class of 1986	Class of 1987	Clas of 1988	Class of 1989
63.7	63.3	60.5	61.2	63.8	66.5	66.0	68.6	68.0	67.2
70.8	69.9	69.4	70.8	73.0	72.3	75.4	74.3	73.1	72.4
42.8	43.0	42.0	40.5	39.2	42.8	45.1	44.4	48.4	44.5
74.4	73.8	70.3	72.2	73.9	73.7	76.2	74.2	76.4	74.4
9.4	11.5	11.7	13.0	14.0	13.0	12.2	11.7	12.3	13.5
23.3	22.4	24.1	22.4	19.2	22.8	21.5	21.0	20.2	23.1

Cancer

Cancer, the second leading cause of death in the U.S., now kills nearly half a million people annually. Some 30% of all cancer deaths—and 90% of lung cancer deaths—are attributable to cigarette smoking. "Reducing the Health Consequences of Smoking," a 1989 report by the U.S. Surgeon General, reviews current statistics on smoking and cancer, among them:

> *To date, 43 chemicals in tobacco smoke have been determined to be cancer-causing (carcinogenic).*
>
> *The latest data show that the average male smoker is 22 times more likely to die from lung cancer than a non-smoker.*
>
> *In 1986 lung cancer caught up with breast cancer as the leading cause of cancer death in women. Women smokers' relative risk of lung cancer has increased by a factor of more than four since the early 1960s, and reflects the growth in the number of women who smoke.*

Smoking and other forms of tobacco use are also associated with cancers of the bladder, pancreas, kidney, larynx, mouth, esophagus, and, it was recently found, the uterine cervix.

Strokes

More than two million people suffer strokes in the United States annually. In 1986, 148,000 people died from strokes, the third leading cause of death in this country. Again, cigarette smoking is believed to be a major cause of strokes.

Chronic Obstructive Pulmonary Diseases

Smoking has been identified as the major cause of chronic obstructive pulmonary diseases, the fifth leading cause of death in the United States. In recent years some 80,000 deaths have occurred annually from COPD, and experts estimate that smoking is responsible for 80 to 90% of those fatalities. The risk of heavy smokers developing chronic bronchitis and emphysema is believed to be as much as 30 times greater than for nonsmokers.

Conclusions of the
U.S. Surgeon General's Report, 1989

Twenty-five years have elapsed since publication of the landmark report of the Surgeon General's Advisory Committee on Smoking and Health. By any measure, these 25 years have witnessed dramatic changes in attitudes toward and use of tobacco in the United States. The health consequences of tobacco use will be with us for many years to come, but those consequences have been greatly reduced by the social revolution that has occurred during this period with regard to smoking.

Since 1964, substantial changes have occurred in scientific knowledge of the health hazards of smoking, in the impact of smoking on mortality, in public knowledge of the dangers of smoking, in the prevalence of smoking and using other forms of tobacco, in the availability of programs to help smokers quit, and in the number of policies that encourage nonsmoking behavior and protect nonsmokers from exposure to environmental tobacco smoke. These changes and other significant developments, as well as the overall impact of the nation's antismoking activities, are reviewed in detail in individual chapters of the report. Based on this review, five major conclusions of the entire report were reached. The first two conclusions highlight important gains in preventing smoking and smoking-related disease in the United States. The last three conclusions emphasize sources of continuing concern and remaining challenges. The conclusions are:

1. **The prevalence of smoking among adults decreased from 40 percent in 1965 to 29 percent in 1987. Nearly half of all living adults who ever smoked have quit.**
2. **Between 1964 and 1985, approximately three-quarters of a million smoking-related deaths were avoided or postponed as a result of decisions to quit smoking or not to start. Each of these avoided or postponed deaths represented an average gain in life expectancy of two decades.**
3. **The prevalence of smoking remains higher among blacks, blue-collar workers, and less educated persons than in the overall population. The decline in smoking has been substantially slower among women than among men.**
4. **Smoking begins primarily during childhood and adolescence. The age of initiation has fallen over time, particularly among females. Smoking among high school seniors leveled off from 1980 through 1987 after previous years of decline.**
5. **Smoking is responsible for more than one of every six deaths in the United States. Smoking remains the single most important preventable cause of death in our society.**

Source: U.S. Surgeon General, "Reducing the Health Consequences of Smoking: 25 Years of Progress," 1989.

Smokeless Tobacco: A New Epidemic?

Smokeless forms of tobacco include chewing tobacco, snuff (a powdered variety of chewing tobacco) and most recently a synthetic cigarette which is not lit but only "puffed" on. Until recently such forms of tobacco were used primarily by a relatively small percentage of the U.S. population located mainly in the southeastern states. However, in recent years smokeless tobacco sales have been increasing at a rate of better than 10 percent per year, and evidence suggests that many of the new users are in their teens and younger. In particular, many male teenagers, hearing of the dangers of smoking and influenced by the example of well-known athletes, have turned to smokeless tobacco products. From 1970 to 1986, males between the ages of 17 and 19 increased their use of snuff fifteenfold and their chewing tobacco use fourfold.

Smokeless tobacco can cause oral cancer and other forms of mouth disease, and like cigarettes, can result in dependence. In addition, the use of smokeless tobacco may produce blood nicotine levels similar to those caused by cigarette smoking. Advertising that uses popular athletes and entertainers to promote these products is considered partly responsible for this most recent "epidemic," as are packaging and labeling with specific youth appeal. Ironically, increased awareness of the health hazards of cigarette smoking may also be a factor in the turn to smokeless tobacco products by people who are unaware of their hazards.

Although smokeless tobacco's popularity tapered off slightly in the late 1980s following the highly publicized death of Sean Marsee, an Oklahoma high school track star who used smokeless tobacco, these products are again increasing in popularity. Doctors who specialize in treating the disfiguring oral cancer linked to chewing tobacco believe that young boys are influenced by advertisements for the product that appear in publications about sports and outdoor activities.

It appears that a growing number of young people, influenced by the example of well-known athletes, are using smokeless tobacco. The result can be oral cancer and other types of mouth disease.

In addition, it has been found that pregnant women who smoke are less likely to give birth to healthy infants. Pregnant teenagers are especially at risk if they smoke and face an increased incidence of miscarriage or stillbirth and a greater chance of having a premature and/or low-birthweight infant. Studies have also shown that the children of parents who smoke are sick more often; they have more colds and much higher rates of bronchitis and pneumonia. Other studies have suggested that nonsmoking wives who live with heavy smokers are twice as likely to die of lung cancer as the wives of men who do not smoke and are also three times more likely to suffer heart attacks. The potentially dangerous effects of "passive smoking" have led to increased restrictions on smoking on trains, planes, and other forms of public transportation; in restaurants; in office buildings; and in other public places where large numbers of people gather.

It should be noted that cigarette smoking causes economic loss as well as human suffering. According to the U.S. Public Health Service, smoking accounts for $22 billion in medical costs each year and another $43 billion in lost production. Medicare and Medicaid alone pay out at least $4.2 billion annually to care for those who are ill from cigarette-related diseases.

The Benefits of Smoking

Scientific studies of the effects of cigarette smoking reveal that many of the early claims about the benefits of tobacco, as well as the above described dangers, were not groundless. Some previously suggested benefits were based on valid observations. Smoking is a convenient way for people to help regulate their mood and feelings, and it has some utility in the control of body weight.

When possible, or useful, benefits of a drug are considered we must address the issues of "safety" and "efficacy." That is, the drug must be proven safe in the form that it will be marketed, and the drug must be proven effective in scientific tests. These are key factors used by the Food and Drug Administration (FDA) when new drugs or food additives are considered for approval.

The nicotine delivered by cigarette smoking clearly has some therapeutic effects. On the other hand, the overwhelming evidence that cigarette smoke also causes a wide range of damaging effects would prevent tobacco from being approved for therapeutic purposes. However, to understand why so many people smoke, why treatment of smokers is difficult, and why most people who quit smoking soon resume the habit, it is important to understand some of the benefits gained by smoking.

As a means of *self-medication* tobacco is very convenient. It is also legal, relatively inexpensive, and readily available. It may be consumed nearly anywhere with minimal social stigma, the amount (dose) can be precisely controlled, and it is in a convenient delivery system. In the form of tobacco smoke, nicotine can be delivered to the bloodstream within seconds of smoke inhalation. The nicotine dose has ex-

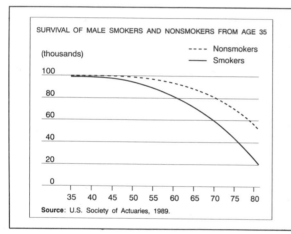

SURVIVAL OF MALE SMOKERS AND NONSMOKERS FROM AGE 35

(thousands)

---- Nonsmokers
—— Smokers

Source: U.S. Society of Actuaries, 1989.

Figure 6. *Compares the survival of male smokers and nonsmokers. Ten percent of smokers die before they reach age 55, compared to only 4% of nonsmokers. By age 65, 28% of the smokers are dead, but only 11% of the nonsmokers. By age 75, 57% of smokers are dead, as compared to 30% of nonsmokers.*

When the Surgeon General's warning first appeared on cigarette packages in 1964, smoking declined immediately throughout the United States. Even stronger warnings on packages took effect in 1985.

tremely rapid and controllable effects. In fact early researchers dubbed this system "finger-tip control of dose," to
emphasize the puff by puff precision with which a smoker
can regulate his or her nicotine dose level.

Once in the bloodstream, nicotine has immediate effects
on many hormones. Two, in particular, have been studied in
detail: epinephrine (or adrenaline) and norepinephrine
(noradrenaline). Epinephrine is released into the bloodstream
when people are anxious, stressed, or bored. Norepinephrine is released during certain kinds of heightened arousal
caused by excitement, exercise, antidepressant drugs, sex,
many drugs of abuse, and nicotine. There is some evidence
that cigarette smokers can use nicotine to adjust their levels
of norepinephrine and thus self-regulate their own moods
and emotional states. It should come as no surprise that two
of the most commonly given reasons for smoking are for
"stimulation" and "reduction of stress."

Studies show that nicotine has some therapeutic effects
probably related to these biochemical effects. In one study
the effects of nicotine on aggression were measured. Two
research subjects were seated in separate rooms in which
they worked at a computer game to earn money. One subject would periodically lose points (and money). This loss
was thought, by the subject, to be due to the deliberate
actions of the other. When this happened the person who
lost the money could either subtract points from the other
person or blast the subject with a loud noise. Before the test
the subject was given a cigarette to smoke. The cigarettes
delivered different doses of nicotine. The higher the dose of

In Japan 75% of all adult males smoke. The government-owned industry sold over 300 billion cigarettes in 1982, up 2.7 billion from 1981.

nicotine that the "angry subject" smoked, the less likely he was to punish the other subject.

Other studies have verified that nicotine can indeed reduce anxiety and make people more tolerant of stressful events and distractions. Together, these studies support claims by many smokers that cigarettes help them to deal more effectively with the daily stresses in life.

Tests have also shown that impairment in performance occurs when the tobacco user is deprived of nicotine. In particular, tasks involving *vigilance* and *concentration,* such as watching a computer screen, are impaired during nicotine deprivation and can be restored by readministration of tobacco or by therapeutic use of nicotine gum. Nicotine does

A Barnum & Bailey's Circus trainer poses in 1913 with "Old Joe," the camel whose likeness still appears on packages of Camel cigarettes.

not, however, improve performance beyond the user's normal capabilities—in other words, it does not help the user to perform tasks better than he would have had he never used tobacco.

Many people claim that they smoke as a means of *weight control,* and there is evidence that nicotine and smoking do facilitate this in at least four ways. First, nicotine decreases the efficiency with which the body extracts energy from food. Thus more food is eliminated without being converted to fat or muscle. Second, nicotine reduces the appetite for foods containing simple carbohydrates (sweets). Third, smoking reduces the eating that often occurs as a response to stress. Fourth, nicotine increases general body metabolism. As a result of these effects, cigarette smokers tend to weigh less than nonsmokers.

Roughly one-third of the people who quit smoking gain weight. Antismoking enthusiasts point out that this is not so bad since two-thirds of the people who quit smoking do not gain weight. But they miss an important point. That is that weight gain is a possible undesirable side effect to quitting smoking. For any other type of therapy, a side effect that occurred in one-third of those treated would be considered worthy of serious attention.

It should be apparent that the effects of smoking are many and varied, but in the long run, smoking is hazardous to health and shortens the expected life span. Understanding the benefits, however, can provide insights as to why some people smoke. More importantly, it should be clear that removing the benefits of smoking may lead to side effects which must be addressed. Otherwise treatment will not be successful.

An American Cancer Society poster points out that smoking during pregnancy can have adverse effects on the unborn child. Associated problems include retarded growth, spontaneous abortion, and malformation. Until adolescence, children of mothers who smoke remain several months behind children of nonsmokers in reading and math.

CHAPTER 4

SMOKING, WOMEN, AND PREGNANCY

ʟ

The increase in cigarette smoking among women is a more recent phenomenon than it is for men. Figure 7 shows trends in the incidence of smoking among the sexes from 1923 through 1980 in a Midwestern metropolitan area. This survey is a good indication of smoking trends in the rest of the United States as well. Men smokers outnumbered women smokers by about 2 to 1 until the 1950s and 1960s. At that time reports linking cigarette smoking and cancer began to appear in the media.

These reports were thought to be a critical factor in reducing the incidence of smoking in men, particularly in men who were better educated and had higher incomes. At about the same time women were becoming more active in social roles and fields of work that had been dominated by men. Women began to seek and achieve higher levels of education and employment, and to compete with men to an unprecedented degree in nearly all aspects of life. Among women, the higher incidence of cigarette smoking occurred in those with higher education and income levels.

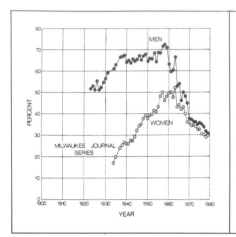

Figure 7. *These data were collected in annual surveys of residents in the Milwaukee area by the* Milwaukee Journal. *The graph shows the percentage of men and women who reported smoking cigarettes during each of the years indicated on the horizontal axis. The substantial decline in the incidence of smoking after 1960 is evidence of increasing public awareness of the dangers of smoking.*

Cigarette manufacturers have kept abreast of social and cultural changes throughout the decades. Advertisers in the 1920s and 1930s targeted certain ads for women. Chesterfield advertisements in 1926 had young women pleading to young men smokers "Blow some my way." About this same time ads for Marlboro cigarettes emphasized that "women, when they smoke at all, quickly develop discriminating taste." Lucky Strike ads urged women to "reach for a Lucky instead of a sweet."

New cigarettes were also developed with potential female consumers in mind. Lorillard was the first major manufacturer to show women smoking, in an ad for Helmar cigarettes in 1919. A more recent product, Satin, was reported to have an advertising and marketing budget of $75 million for 1983. The cigarette was a standard 100 millimeter cigarette, but its appearance (shiny satiny paper) and the attractive package were specifically targeted at supposedly progressive young women. Other companies took similar approaches. The extensive campaign for Virginia Slims in the 1970s emphasized these cigarettes as the ones for high-achieving, competitive, and liberated women: "You've come a long way, baby."

The Effects of Smoking on Women

It is a myth that women are immune to the adverse effects of cigarette smoking. In both men and women, the prevalence of all major types of chronic respiratory diseases (bronchitis, asthma, impaired breathing) is directly related to the level of smoking. Similarly, as women have begun to smoke more, their incidence of lung cancer has shown a steady rise. Additionally, women who take birth control pills are at greater risk of cardiovascular disease if they smoke cigarettes.

There is even evidence that cigarette smoking affects some aspects of the sexuality of women. Smoking more than half a pack of cigarettes per day is associated with a higher incidence of infertility. Irregular menstrual cycles are more prevalent among women who smoke cigarettes. The number of years of potential fertility is also reduced. Menopause occurs earlier among women who smoke.

The Effects of Smoking on Offspring

The effects of cigarette smoking on pregnancy, birth weight, and infant health have been studied extensively. Figure 8 shows the effects of cigarette smoking on birth weight and several other aspects of pregnancy. Babies born to women who smoke are an average of 200 grams (about 7 ounces) lighter than babies born to nonsmokers. This is important since birth weight is an excellent predictor of infant health. It appears that this retarded growth is caused by hypoxia or decreased oxygen available to the fetus. This is partly due to the carbon monoxide (CO) delivered by smoke inhalation.

Another effect of cigarette smoking during pregnancy is the increased likelihood of spontaneous abortions. In fact,

Until the United States government banned cigarette advertising on radio and television in 1971, tobacco companies traditionally sponsored shows.

the risk is almost double for women who smoke. Smoking also increases the risk of congenital malformations. This, like so many other effects of smoking, is directly related to the amount of smoking. Levels of smoking are also associated with a variety of other complications during pregnancy and labor. These include increased risk of bleeding and premature rupture of membranes. Finally, there is a clear relationship between smoking during pregnancy and the occurrence of the sudden infant death syndrome (SIDS).

Babies born to cigarette smokers develop more slowly throughout childhood than babies born to nonsmokers. They are more likely to have neurological (brain function) disorders, psychological abnormalities, and lower intelligence scores. Until adolescence, children of mothers who smoke 10 or more cigarettes per day remain about three to five months behind children of nonsmokers in reading, mathematics, and general ability scores. Cigarette smoking during pregnancy is also a significant risk factor for hyperactivity in children.

The mechanisms that underlie the harmful effects of

A turn-of-the-century poem by James Harvey satirizes the smoking habit.

A cigarette ad from the 1920s aims squarely at image-conscious female smokers.

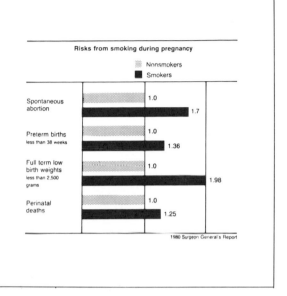

Figure 8. *This chart shows the number of various kinds of pregnancy problems for women who smoke cigarettes compared with such problems in women who do not smoke.*

Risks from smoking during pregnancy

Nonsmokers
Smokers

Spontaneous abortion — 1.0 / 1.7

Preterm births less than 38 weeks — 1.0 / 1.36

Full term low birth weights less than 2,500 grams — 1.0 / 1.98

Perinatal deaths — 1.0 / 1.25

1980 Surgeon General's Report

smoking on the fetus and on infants include hypoxia (oxygen starvation) due to the CO carried by the smoke into the lungs. Nicotine, which crosses the placenta, may raise the fetus' blood pressure and thus induce slowing of the heart. Tar also crosses the placental barrier but harmful effects have not been clearly proven to date. Animal studies have shown that even brief exposure to nicotine can result in structural changes in the nervous system. These findings suggest that the process of becoming vulnerable to nicotine dependence, or even developing some dependence itself, can begin before birth.

All these data make it clear that cigarette smoking is a risk factor for both mother and infant. While smoking does not ensure damage, it is an avoidable risk factor. Since much of the risk is to the unborn child, the mother bears a great responsibility.

A technician at the Louisiana State University School of Medicine lights up a cigarette for a baboon as part of a study investigating hardening of the arteries caused by smoking. Scientists have used animals as experimental subjects in smoking-related tests since the 1950s, when serious research into the medical effects of smoking first began.

CHAPTER 5

THE SCIENTIFIC STUDY OF CIGARETTE SMOKING BEHAVIOR

Until the 1970s there was little effort made to study the behavior of cigarette smoking. This is in sharp contrast to the enormous expenditure in dollars and research hours spent studying the effects of smoking on health. Even the amount of money aimed at trying to convince or help people to quit smoking far surpassed that spent on trying to understand the behavior itself.

Resistance to Study of Cigarette Smoking Behavior

Intensive research into why people smoke did not begin until the late 1970s. There were scientific, political, and practical reasons for this. Until recently, cigarette smoking was held to be a normal voluntary behavior that was a source of pleasure. Advertising campaigns have been careful to develop and perpetuate this view of smoking. Cigarette smoking was presented as a means of building a particular image, perhaps one of sophistication.

In the 1920s and 1930s cigarettes were endorsed by famous athletes and movie stars, who attributed part of their success to smoking a particular brand. Some of the more successful recent images have been those of the rugged Marlboro Man, the dedicated and independent Camel smoker, and the confident, liberated Virginia Slim woman. Another technique commonly used by cigarette advertisers is to associate smoking with the beauties of nature and healthy outdoor living.

Since the scientific study of behavior is usually reserved for actions considered deviant, pathological, or related to mental health or drug problems, cigarette smoking, which was seen as a normal and voluntary behavior, received little study. The view of cigarette smoking as a voluntary pleasurable act was most prevalent from the early 20th century until recently. Thus, it was reasoned, smokers simply needed to be informed that smoking was harmful to their health; then they could weigh the risks and benefits in a rational manner, and presumably, would quit. The other side of the coin is that when a person failed to quit smoking, it was assumed that he or she did not really understand that smoking was harmful, or that the person had a self-destructive personality.

This logic led to antismoking campaigns which showed lung cancer operations. For some people, informing them that smoking was harmful was sufficient to cause them to quit. However, as King James learned in the early 17th

Every 20 seconds for five years until September 1977, the typically rugged man portrayed in this Chicago cigarette advertisement took a puff and blew perfect smoke rings. Tobacco companies spend vast amounts of money on marketing every year.

century, for most people such campaigns were of little lasting value.

The political and economic reasons that delayed the study of cigarette smoking behavior are intertwined. The government subsidizes the tobacco industry in much the same way that it subsidizes many other facets of the agricultural industry. Such subsidies are critical to many state and local economies for which tobacco is a major source of revenue.

For the federal government, money derived from cigarette taxes amounts to billions of dollars per year (see Figure 1). As a result, the government is reluctant to take actions that might impair this revenue. In the past, the government supported the tobacco industry in other ways as well. Until recently the Veterans Administration and the armed services subsidized cigarette sales to military personnel. Many of these practices were established long before there was definitive scientific evidence that smoking was harmful.

In 1959, leaders of the Seventh Day Adventist Church, working in conjunction with physicians, devised a program designed to help smokers quit the habit without the use of drugs. The program, presented in five sessions, concentrates on the physical, mental, social, and spiritual implications of smoking.

From a practical standpoint, studying the behavior of cigarette smoking also represented a considerable challenge. Social, cultural, pharmacologic, and behavioral variables are involved in smoking. The behavior itself is complex. It was not clear which aspects of cigarette smoking were important to measure. The number of cigarettes smoked tells little about how they were smoked or how much cigarette smoke was taken in by the lungs. Number of puffs and volume of smoke inhaled are not easily automated. It was not even clear how to control the dose of smoke given.

The fundamental questions of what aspects of cigarette smoking to measure and how to measure them remained stumbling blocks to cigarette smoking research until recently.

This 1944 magazine advertisement for Camel cigarettes has a patriotic theme: a U.S. Infantry soldier enjoys a well-earned smoke. The ad invites the reader to identify with this G.I. Joe and try "the fighting man's favorite cigarette."

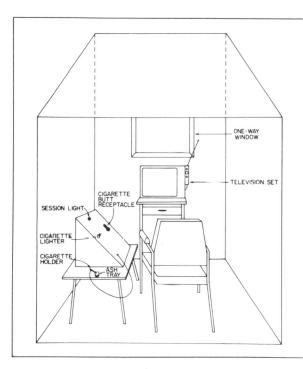

Figure 9. *The instruments in this cigarette smoking test room are wired to a computer in an adjacent room so that every puff on every cigarette is automatically measured.*

Scientific Laboratories and Devices to Study Cigarette Smoking

Before studies of cigarette smoking could proceed, laboratories and devices had to be developed to study the behavior. Laboratories were developed to study smoking at a macro level (where larger aspects of smoking are studied, such as the effect of a cold on the number of cigarettes smoked) and a micro level (which might consider what happens during individual puffs). For the most controlled level of study, test rooms were developed in which volunteers could smoke cigarettes while computers recorded their behavior. One such test room is shown in Figure 9. In this context it was possible to carefully study the behavior of cigarette smoking and to discover what factors controlled it.

The cigarette smokers who served as subjects in these studies did not feel that they smoked differently than normal while in the laboratories. However, it was important to prove this, in other words to validate the system. To do this, portable puff-monitoring systems were developed. These systems take the scientist one step away from the laboratory where events are more controlled, and one step closer to the natural environment of the smoker. These allowed smoking behavior to be monitored by people as they went about their daily living in their usual environments.

Another type of laboratory setting makes less use of automated devices but permits rigorous observation of cigarette-smoking behavior. This is the residential research laboratory where volunteers live in the company of other volunteers for periods of a few days to several months. In such a setting, persons are free to smoke as they choose, but they obtain each cigarette from either cigarette-dispensing machines or from research staff. Patterns of smoking are thus observed 24 hours a day.

Each of these settings is uniquely arranged to measure some aspect of smoking behavior. The strategy of using several systems permits a variety of checks (points of

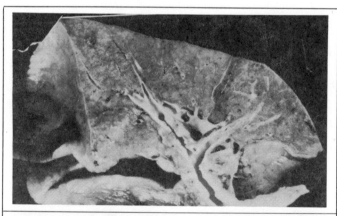

A normal human lung, with blood vessels clearly visible, no discoloration, and no tumorous growths.

A lung taken from a victim of emphysema, exhibiting extreme discoloration and much damaged tissue.

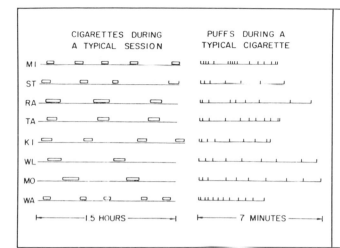

Figure 10. *Typical patterns of smoking cigarettes (left side) and puffing on each cigarette (right side) are shown for each of eight volunteers. In the course of the experiment, smoking emerged as an orderly behavior in which addicts compensated for deprivation by smoking more intensely when a cigarette was finally given.*

validation) to be made. It also provides a much more complete picture of smoking.

Descriptive Studies of Cigarette Smoking Behavior

The first step for scientists using these laboratories and devices was to carefully measure all that happens when people smoke cigarettes. The first studies recorded what times people smoke. It was learned that cigarette smoking was not a random event, but rather an orderly behavior. Figure 10 shows smoking patterns obtained in one laboratory study. Similar patterns were recorded whether by staff observations or by portable puff-monitoring devices. Furthermore, the relationship between number of cigarettes smoked and CO levels was similar whether or not puff-monitoring systems were used. Since each system gave the same result, it was more likely that the results were valid.

Another set of studies showed that puffing and inhaling were also very orderly. As the cigarette grew shorter, puffs tended to become smaller in duration and volume. It also appeared that when strong cigarettes were smoked, many people took smaller puffs but diluted them by taking in greater volumes of air when they inhaled the smoke. Depriving smokers of cigarettes resulted in their smoking more cigarettes and taking larger puffs when they were again permitted to smoke. The findings suggested that cigarette smoking was controlled to a large extent by a drug contained in the tobacco smoke.

The Effect of Different Features of Tobacco on Smoking

When drugs of abuse are studied, the dose, or the amount of drug taken, is a critical factor. The general finding is that larger doses are taken less often than smaller doses, resulting in relatively stable amounts of the drug being taken. This effect is termed "compensation" or "dose-regulation" and occurs in many forms of drug use. For instance, people tend to drink beer in much greater volume than whiskey. With regard to cigarette smoking this issue of compensation has both theoretical and practical implications.

Theoretically, it is important to discover if people regulate cigarette smoke the same way they regulate intake of drugs. If they do, then it is important to discover precisely what they are regulating: Is it the smoke itself or the nicotine?

Practically, it is important because one way of treating smokers is to have them gradually decrease the number of

During the 1930s many celebrities, including star athletes, were contracted by tobacco companies to attribute part of their success to smoking. Despite the many accomplishments of Lou Gehrig, the fact remains that fitness and smoking are rarely compatible.

cigarettes that they smoke. People who can't quit smoking are also encouraged to smoke cigarettes delivering less tar and nicotine on the assumption that they are safer. But smokers who compensate for changes in the strength of their cigarettes by smoking more of them or inhaling more deeply will defeat their purpose.

One study used an automatic puff-monitoring system to measure smoking behavior when subjects were required to smoke through ventilated cigarette holders. (These One Step At A Time® holders were sold to help people quit smoking by gradually cutting back their dose.) The main finding of the study was that the smaller the dose permitted by the filter, the more puffs were taken. These smokers did not smoke many more cigarettes, but with the extra puffs they were able to take in nearly as much smoke. This was proved by the fact that their CO levels were similar regardless of the holder used.

"But, Dad, it's not like when you were a boy.
The smoke in these cigarettes is pulled through tiny, scientific, mentholated air pores,
where thirty-five thousand filters trap the nicotine and tars,
and protect the throat from harmful irritation."

As this cartoon illustrates, cigarette companies began emphasizing the benefits of filters in screening out harmful nicotine and tar, following the release in 1964 of the U.S. Surgeon General's Report on the harmful effects of smoking.

Two other approaches to manipulating dose were changing the number of puffs permitted per cigarette and cutting cigarettes into various lengths. The general finding of these approaches is the same as that in the ventilated holder study. Namely, people compensate for changes in dose and end up taking in about the same amount. In these studies more cigarettes were smoked when the cigarettes were shorter, or when fewer puffs were permitted per cigarette.

These studies showed that people regulate tobacco smoke intake quite well, but was it the smoke or the nicotine contained in the smoke that they were regulating? Until recently uncertainty existed because several studies had shown that if special cigarettes were given in which the smoke was the same in all respects except nicotine level, people did not appear to regulate the number of cigarettes they smoked as well as would be expected. The results of the ventilated study offered a clue as to what might be happening. Namely, smokers might regulate in subtle ways, like number of puffs or the depth of inhalation.

The way to address the issue was to measure the amount of nicotine actually taken into the body. Several such studies were done in the 1980s. These studies showed that the amount of nicotine in the blood had little to do with the amount of nicotine in the cigarette. It appeared that smok-

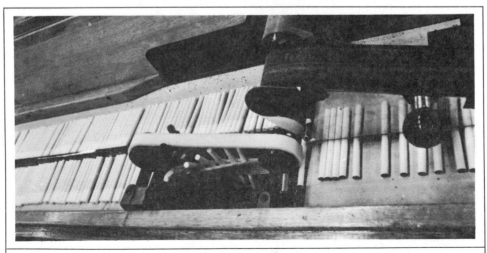

A conveyor belt in a Durham, North Carolina cigarette factory, capable of carrying up to 500 billion cigarettes each year.

ers regulated their smoking behavior so that they obtained a certain nicotine level regardless of the dose in the cigarette.

These studies uncovered another point of similarity between the use of tobacco and other drugs of abuse. The dose of the substance is a critical factor in controlling the pattern of drug intake. The practical implication of this finding is that instructing people to switch to lower tar and nicotine brands of cigarettes may result in a false sense that they have lowered their health risk. The evidence indicates, in fact, that smokers of lower tar and nicotine cigarettes take in nearly as much of the substances contained in smoke as smokers of stronger cigarettes. This also helps explain the poor results of such approaches to quitting as using ventilated holders and switching brands.

Other Drugs and Smoking

Drugs of abuse generally have a variety of behavioral effects. It is important to learn of these effects on cigarette smoking

Blacks predominated in lower-grade process work in the tobacco industry throughout the 19th century. Visible between the two women with brooms, a white foreman supervises production in a Virginia tobacco factory.

for both theoretical and practical reasons. Theoretically, discovering how certain drugs affect cigarette smoking behavior can help us to understand the nature of the behavior itself. From a practical standpoint it is important to learn if drugs can be used to help treat smoking, or if taking drugs will hinder treatment efforts.

One of the first drugs to be carefully studied for its effects on smoking was alcohol. It is commonly claimed by smokers that they smoke more when they drink. The laboratory studies proved this claim objectively and went a few steps further.

First, the studies showed that it wasn't just a social phenomenon. It even occurred when people drank in isolation. Second, people didn't simply light up more often. The measures of puffs taken and CO levels showed that people actually inhaled more. Third, it wasn't simply due to the person's expectation: it happened when the flavor of the beverage was masked, and the stronger the dose of alcohol the more they smoked. It was dependent, however, on past experience of drinking and smoking, because the effect was very weak in light social drinkers, and alcohol may actually decrease smoking in people who hardly drink at all. Seda-

In September 1990, U.S. Surgeon General, Dr. Antonia Novello, announces the results of a new study on the benefits of quitting smoking. The report notes that even when smokers decide to quit at age 50, the former smoker's chances of dying are cut by one-half within 15 years.

tives (pentobarbital) and opioid narcotics (methadone and morphine) also increased cigarette smoking in abusers of these drugs.

One drug of particular interest was *d*-amphetamine. Amphetamines have been prescribed to help people quit smoking on the theory that cigarettes are stimulants and that substituting one for another should reduce smoking. However, the theory is oversimplified (smoking can both stimulate and relax), and the treatment has not worked. Figure 11 shows that when given *d*-amphetamine, people actually smoked more. This study measured several factors and showed that people smoked more cigarettes, took more puffs, achieved higher levels of smoke intake (CO), smoked the cigarettes down further, and felt better while smoking.

The findings regarding *d*-amphetamine suggest an interesting theory. Namely, drugs that produce positive feelings (euphoria) may increase cigarette smoking. Certain drugs such as amphetamine appear to be somewhat more univer-

Actress Brooke Shields lends her support to a campaign by the National Council on Alcoholism. Researchers have discovered that experimental subjects smoke more intensely and more frequently when drinking alcoholic beverages.

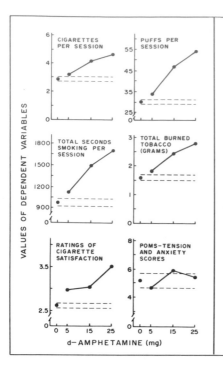

Figure 11. *Eight volunteers were tested in 90-minute sessions during which several aspects of cigarette smoking (shown on the vertical axis) were measured. Two hours before each session they were given a capsule that contained either a single dose of amphetamine or a placebo (shown on the horizontal axis.) The zero on the horizontal axis stands for the placebo, and thus on each graph the point above the zero and the area within the dashed lines represents the normal, drugless state. Points falling outside the dashed lines indicate that the effect of the amphetamine was significantly different from that of the placebo. The scale in the right bottom panel shows that the effect was not due to nervousness.*

sal euphoriants and may affect most smokers in the same way. Other drugs such as alcohol may only be euphoriants in certain individuals and may only increase smoking in these people. Caffeine has relatively little effect on smoking. This is probably because at the same time that caffeine provides some stimulation and euphoria, it can also make people more nervous. These findings on the effects of psychoactive drugs have important treatment implications. One is that many of the drugs that an individual commonly uses while smoking cigarettes should be avoided while he or she is trying to quit smoking cigarettes.

The findings of the above studies led to a hypothesis that drugs that blocked the effects of nicotine in the body might alter cigarette smoking regardless of their effect on mood. Therefore, it was predicted that treatment with nicotine not in tobacco form would reduce cigarette smoking. This is because, even though nicotine has some euphoriant effects, it would be like giving a person much stronger cigarettes and he or she should compensate by smoking less. In several studies volunteers were given nicotine in either an intravenous form or in the form of a chewing gum. The results were consistent with the hypothesis; that is, people smoked less.

A drug that diminishes the effects of nicotine should increase cigarette smoking as the smoker attempts to compensate. A study was done in which an antihypertensive medication, mecamylamine, was given in capsule form. According to the theory, if the dose of nicotine is suddenly reduced because it has been blocked, the person should smoke more. The results were consistent with the hypothesis; that is, people smoked more. Theoretically, if such a blocker were given continuously, the person would ultimately quit since he or she simply wouldn't be getting any pleasure out of smoking. Preliminary testing proved this to be true. These results suggest that such medication might ultimately help people to quit smoking if repeated on a daily basis.

THE HISTORY AND MYSTERY OF TOBACCO.

TO what extent active stimulants are necessary for the health of the body and the development of the intellect, affords a subject of speculation which, it seems, will never be brought to a satisfactory conclusion. Speaking without referring to the experience of all ages, we would say that, beyond a sufficiency of wholesome food, nothing more was necessary to sustain the human body in its greatest perfection; yet it is notorious that, from the earliest ages and among all peoples, the custom has prevailed of using a thousand substances, evidently for no other purpose than to give unnatural acceleration to the system; and thus, through the body, add im-

Entered according to Act of Congress, in the year 1855, by Harper and Brothers, in the Clerk's Office of the District Court for the Southern District of New York.
VOL. XI—No. 61.—A

A 19th-century tract against smoking raises a question debated to this day: why do human beings take stimulants when such substances are not essential to sustaining life?

A young girl lights a cigarette. As recently as the 1960s male smokers outnumbered females 2 to 1, but today more females than males smoke, despite the fact that overall smoking rates for the total population have declined. According to the Surgeon General's report, 1989, smoking begins primarily during adolescence, with females smoking earlier than males.

CHAPTER 6

EFFECTS OF NICOTINE

*T*he importance of nicotine in cigarette smoking has to do with its effects on the brain. This is not a new revelation. Centuries ago it was recognized that tobacco produced "mind altering" effects, that it was a "food for the brain." We now know that many of these effects are not unique to nicotine but are shared with most other drugs of abuse. To understand why people smoke cigarettes when they know it is harmful to their health, it will help to know what nicotine does to the brain.

How Nicotine Acts in the Brain

To affect the brain, a drug must have physical properties that permit it to pass through the blood–brain barrier (a barrier that helps protect the brain from foreign substances). The nicotine molecule has ideal physical features that allow it to penetrate the brain easily. When nicotine is inhaled into the lungs, the arterial bloodstream picks it up and carries it to the brain within 10 seconds.

Once in the brain, nicotine molecules work like keys opening locks. The locks, called "receptors," are located on nerve cells of the brain and connect it to muscle tissues and the various organs of the body. Drugs that act as keys are called "transmitters" since they help to send information from one part of the body to another.

Figure 12 shows a transmitting and receiving station (nerve cell). Receptors located on the receiving end (dendrites) of the nerve cell transmit information by an electric impulse through the axon. At the end of the axon

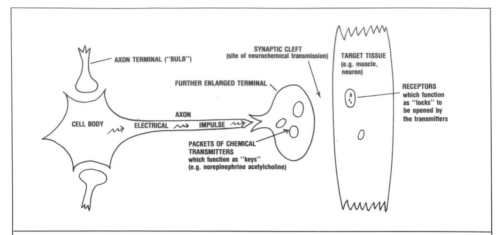

Figure 12. *The drawing represents a "typical" neurotransmitting system. The neuron receives a chemical impulse from its dendrites. This is converted to an electrical impulse that passes down the axon to the nerve terminal. There, a small amount of a chemical transmitter is released. The transmitter diffuses across the synaptic cleft where it bonds to a receptor. Most drugs act at this synaptic cleft. For instance, cocaine and nicotine both cause some extra norepinephrine to be released from terminals that use norepinephrine as transmitters.*

terminal a small amount of the body's own chemical transmitter (for instance, adrenaline) is released. The transmitter crosses the gap (synaptic cleft) where it has its effect on the target organ; the target may be another nerve cell which is fired, an organ such as the heart which is speeded up, or a muscle which is stimulated.

Nicotine has complicated actions because the receptors activated by it are located throughout the body. In fact, much of the mapping of the nervous system was done by placing nicotine at various places and charting the effects. This is much like mapping hidden currents in a lake and river system by placing dyes in the water and tracing their course.

Places that were activated by nicotine were called nicotinic receptors. Some of these receptors are located on the adrenal gland. Activation of them results in the release of adrenaline and noradrenaline into the body. These chemicals increase heart rate and blood pressure, heighten arousal, and cause feelings of excitement. Cocaine also results in heightened adrenal activity and has effects on the brain similar to those of nicotine.

A drug that activates a receptor is called an "agonist." Other substances that share some of the critical physical properties of the agonist may substitute for the agonist— much as a master key can substitute for a particular key. For instance, codeine, morphine, and heroin can all substitute for one another. It merely requires adjustment of the dose to get the same effect.

Nicotine acts by substituting for the body's own chemical (acetylcholine) at nicotinic receptors. Lobeline, contained in some over-the-counter aids to help quit smoking, partially substitutes for nicotine. Nicotine in the form of a chewing gum (Nicorette) substitutes very well for nicotine in tobacco smoke. Amphetamine appears to substitute for some of nicotine's effects. Both drugs increase the activity of epinephrine and norepinephrine.

A drug that blocks the effect of an agonist drug is called an antagonist. Some drugs act as antagonists by stimulating other systems (functional antagonists). For instance, giving coffee to a person who is intoxicated by alcohol produces the well-known "wide awake drunk" phenomenon. Other drugs act as antagonists by occupying the receptor without having any effects of their own, thus preventing the agonist drug from working (competitive antagonists). This is like

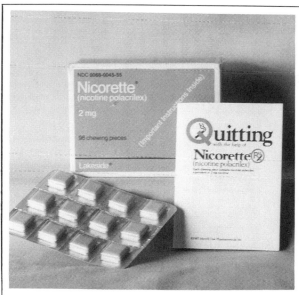

Nicorette, a chewing gum containing nicotine, was developed in Sweden and introduced in the United States in 1984 as a substitute for cigarettes. While acknowledging that this smokeless form of nicotine may be less harmful to the body than inhaling fumes from burning tobacco, experts warn against dependence on any form of the drug.

filling a keyhole with wax, thus preventing the key from entering. An antagonist for nicotine is called mecamylamine. Mecamylamine and similar drugs (ganglionic blockers) are used to treat patients with high blood pressure. They are currently being tested as a possible help in quitting smoking since they block the pleasurable effects of nicotine.

Physiological Dependence on Tobacco

Physiological dependence on a drug occurs when nerve cells have adapted so well to the drug that they require the drug for normal functioning. A person is physiologically dependent on a drug when sudden drug abstinence is followed by withdrawal signs. These may be thought of as "rebound" effects. For example, if heart rate is increased by taking a drug, then abstinence may result in decreased heart rate for a few days. Similarly, feelings of contentment that accompany cigarette smoking may be replaced by irritability, and weight loss may be replaced with weight gain. Many other drugs that are abused produce rebound effects. For example, the constipation that normally accompanies narcotic use is replaced by diarrhea; the muscular relaxation produced by alcohol and sedative drugs is replaced by tremors and even convulsions. When drugs produce extensive and possibly life-threatening rebound syndromes, as is the

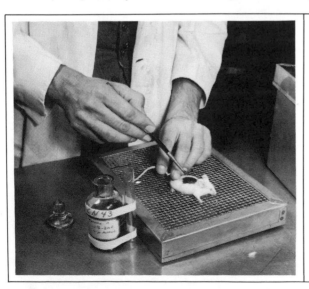

Mice are regularly chosen for experimental laboratory tests to determine the various effects of tar and nicotine on living tissue. They are frequently chosen for experiments because they are easy to handle and house, and also because they reproduce rapidly, thus providing a constant supply of laboratory subjects.

case with opiates and sedatives, the drug is said to be capable of producing physiological dependence.

The presence of a distinct withdrawal syndrome accompanying nicotine deprivation was officially recognized by the American Psychiatric Association in its *Diagnostic and Statistical Manual* in 1980. About one-third of the people who quit smoking report significant withdrawal effects or physical discomfort, although these are usually not as severe as those associated with drugs such as morphine or alcohol.

In certain respects addiction to nicotine is more like cocaine dependence than morphine or alcohol dependence. With all four drugs, behavior becomes compulsive because of the physical effects of the drug on the brain. However, the symptoms of withdrawal from nicotine and cocaine are not as readily detected and were not widely recognized until recently.

A common misconception is that drug abuse is the same as physical dependence; it is not. Physical dependence need not occur to produce drug abuse, and if it does occur, it does not ensure that drug abuse will occur. For instance, many people periodically abuse alcohol or narcotics without ever becoming physically dependent. Also, there are drugs with an enormously high potential for abuse, such as cocaine and amphetamine, which are often used in patterns that do not produce a clear-cut syndrome of withdrawal.

Neatly bundled cigarette butts are displayed in a Tokyo station by a Japanese citizens group to discourage commuters from littering.

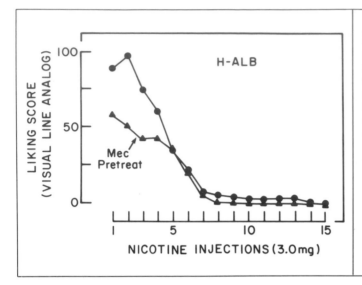

Figure 13. *The decline in the liking score demonstrates the tolerance to nicotine which the body acquires during repeated intake of the drug.*

On the other side of the coin, a person may become physically dependent on a drug and yet never seek that drug or learn to abuse that drug. This happens frequently when people are given narcotics for pain relief. Only a very small portion of these people abuse the drug or continue to take it when it is no longer medically required.

Tobacco Tolerance

As cigarettes are smoked throughout the day, the effects become smaller and smaller. Often smokers assert that their first cigarettes of the day are the "best tasting," with the rest being smoked out of habit or need. When repeated use of a drug results in smaller and smaller responses, the phenomenon is called "tolerance." In other words, it takes more of the drug to produce the same effect. This explains why the first few cigarettes of the day cause the greatest increase in heart rate and blood pressure, the greatest decrease in the knee reflex, and the strongest psychological effect. A laboratory demonstration of this effect is shown in Figure 13. Repeated injections of nicotine produced progressively smaller psychological effects on the subject.

Tolerance develops when the body becomes more efficient at detoxifying and eliminating the drug (metabolic tolerance). People who have smoked for many years eliminate nicotine from their systems much more quickly than nonsmokers. Tolerance also occurs when the nerve cells become less responsive to nicotine doses.

With nicotine, much of this tolerance is lost overnight and quickly regained when smoking resumes the next day. Finally, tolerance can also occur by means of behavioral adjustments as the person learns to compensate for any disruptive effects of the drug. All of these forms of tolerance occur with nicotine, and all appear to be critical in allowing people to quickly learn to enjoy smoking and to get over sickness that often occurs when people first try it. In addition, it is possible that the process of nicotine tolerance may begin in children who are exposed to tobacco smoke by their parents. Studies have shown that children of smokers inhale significant amounts of nicotine. Of course, the fetus of a pregnant smoker is exposed to large quantities of nicotine, too.

A stylish gentleman's smoking excites the curiosity of local children in a 19th-century German cartoon. The introduction of tobacco caused much controversy in Germany, where many officials considered it a threat to civilized behavior.

A 1953 advertisement for Lucky Strike seeks to place the brand firmly in the context of all-American wholesomeness (outdoor living, barbecue time, a happy couple). Creative advertising emphasizing the social pleasures of smoking continues to this day.

CHAPTER 7

ARE TOBACCO AND NICOTINE ADDICTING?

A recurrent theory in the history of the use of tobacco is that its use is compulsive and similar to other forms of drug abuse. In the Americas the inability of the Indians to abstain from tobacco raised problems for the Catholic Church. The Indians insisted on smoking even in church, as they had been accustomed to doing in their own places of worship. In 1575 a church council issued an order forbidding the use of tobacco in churches throughout the whole of Spanish America.

Soon, however, the missionary priests themselves were using tobacco so frequently that it was necessary to make laws to prevent even them from using tobacco during worship. In Italy, as well, there were such problems. One incident in particular is telling. A priest who was celebrating the mass took a pinch of snuff just after receiving the Holy Communion. He began to sneeze, and vomited the Blessed Sacrament onto the altar in front of the congregation.

As tobacco smoking spread through England the demand often exceeded the supply, and prices soared. London tobacco shops were equipped with balances; the buyer placed silver coins in one pan and might receive in the other pan, ounce for ounce, only as much tobacco as he gave silver. The high price, however, did not curb demand.

In 1610 an English observer noted: "Many a young nobleman's estate is altogether spent and scattered to nothing in smoke. This befalls in a shameful and beastly fashion, in that a man's estate runs out through his nose, and he wastes whole days, even years, in drinking of tobacco; men smoke even in bed." A failure of a local tobacco crop could make men desperate. Sailors approaching the island of Nias in the Malay Archipelago were greeted with cries: "Tobacco, strong tobacco. We die, sir, if we have no tobacco!"

The addicting nature of tobacco was noted at about the same time by Sir Francis Bacon, who wrote: "The use of tobacco is growing greatly and conquers men with a certain secret pleasure, so that those who have once become accustomed thereto can later hardly be restrained therefrom."

The invention of cigarette-rolling machines made the availability of tobacco even easier and was seen by some as an additional cause for alarm. The following editorial in the *Boston Medical and Surgical Journal* (1882) states the case well:

> *Our greatest danger now seems to be from an excess of cigarette smoking. The numbers of young men who smoke cigarettes is almost startling. It is not only students, but even school boys in their teens, who vigorously and openly indulge in this dangerous habit. . . . A little cigarette, filled with mild tobacco which lasts for only a few minutes, appears harmless enough. But the very ease at which*

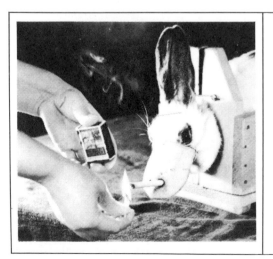

By 1967, when this photograph was taken, this rabbit had smoked nine cigarettes a day for five years in the cause of cancer research at the Institute of Oncology in the Soviet city of Tbilisi. The rabbit came to exhibit a classic symptom of nicotine addiction, only becoming nervous when technicians were late with the cigarette.

these bits of paper can be lighted and smoked adds considerably to the tendency to indulge to excess. . . . One of the pernicious fashions connected with cigarette smoking is "inhaling." The ideal cigarette smoker is never so happy as when he inhales the smoke, holds it in his air passage for some time, and then blows it out in volume through nose and mouth. If he realized that "the smoker who draws the greatest amount of smoke and keeps it longest in contact with the living membrane of the air passages undoubtedly takes the largest dose of the oil," he might at least endeavor to modify his smoking in this respect. . . . These are dangers superadded to those attendant on the ordinary use of tobacco [other than machine rolled cigarettes], and should be considered by all medical men.

Current Controversy and Theory

It has long been suspected that tobacco is a drug of high abuse potential and that nicotine is the critical substance. Some held that nicotine was simply a toxin delivered by tobacco smoke and that it had little to do with behavior. Others felt that nicotine actually was a noxious or aversive

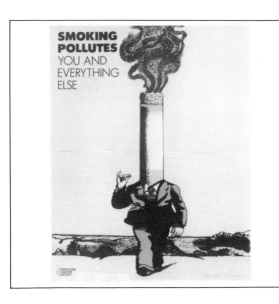

SMOKING POLLUTES YOU AND EVERYTHING ELSE

An American Cancer Society poster highlights the fact that smoking not only endangers the health of the smoker, but also reduces the quality of the smoker's immediate environment.

element in cigarette smoke and limited how much a person smoked. Still others were convinced that nicotine itself was an abusable drug and the key to compulsive tobacco use.

Which theory was correct had implications for the understanding and treatment of cigarette smoking, as well as for government policy. For instance, should treatment be modeled after that used for people who bite their fingernails, or after that for people who abuse drugs? Should the government support the tobacco industry as it does other forms of the agriculture industry, or should it regulate tobacco as a drug?

These issues have been debated in both houses of Congress. One senator stated that cigarette smoking should be considered in the same category as "potato chip consumption." Expert witnesses were found to support every conceivable position. One testified that the high incidence of cancer in cigarette smokers was due to personality traits in smokers. Representatives of the tobacco lobby argued that any regulation on cigarette smoking would be an attack on the free will of Americans to engage in voluntary pleasurable acts.

In 1988 the U.S. Surgeon General came to the following conclusions about tobacco use: (1) Cigarettes and other forms of tobacco are addicting. (2) Nicotine is the drug in tobacco that causes addiction. (3) The pharmacological and behavioral processes that determine tobacco addiction are similar to those that determine addiction to drugs such as heroin and cocaine. Earlier, the director of the National In-

A Michigan resident dons a gas mask during a football game to protest smoking in public places. The controversy over smoking in public involves two powerful lobby groups: the tobacco industry and the public health associations.

stitute on Drug Abuse (NIDA) also concluded that cigarette smoking was a form of drug abuse. He insisted that cigarette smoking is not a voluntary pleasure but rather a compulsively driven behavior. Some of the evidence that led both the Surgeon General and NIDA to take this position is summarized in the following pages.

Similarities in the Use of Tobacco and Known Drugs of Abuse

The most obvious similarity is that, by and large, non-nutritive plant products are not widely consumed unless they contain a drug that affects the way people think or feel (psycho-activity). These substances are used in such a way that the drug gets into the bloodstream and ultimately to the brain. Opium poppies are reduced and refined to yield the potent extract morphine; further processing results in heroin. Cocaine is extracted from coca leaves and processed in ways that maximize its effects.

Nicotine is made available, following an elaborate process of harvesting and manufacturing, in a very convenient and effective delivery system—the tobacco cigarette. Nicotine reaches the brain even more efficiently when inhaled in tobacco smoke than when given intravenously. It is well absorbed through the thin membranes of the mouth and

Irene Parodi of Fremont, California, smiles after a landmark case in which a Circuit Court of Appeals awarded her $20,000 in disability pay because she developed bronchitis after being transferred to an office where most of her co-workers smoked. Recent evidence indicates that even nonsmokers can suffer from being in the presence of smokers.

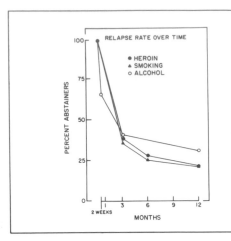

Figure 14. *A survey from the early 1960s shows the rates of relapse for three drugs of abuse.*

nose and is therefore well absorbed when taken in the form of chewing tobacco or snuff.

Developing patterns of the behavior of tobacco or drug use are also similar. With tobacco, alcohol, heroin, and many other drugs, initial use (experimentation) is usually with the support or even encouragement of friends or relatives. One of the best predictors of smoking, in fact, is that both parents are cigarette smokers. Similarly, a common finding among people who have recently quit smoking is that either their friends or relatives have also quit. These patterns of peer and social pressure are shared with most drugs of abuse.

Patterns of use also are similar. As with the opiate drugs (morphine and heroin), initial buildup of tolerance leads to increased use, which then remains stable over long periods of time. Taking the drug away for short periods results in discomfort and an increased desire to take the drug.

In a study and treatment program for smokers, heroin addicts, and alcoholics, people were required to abstain from the use of the drug. Over the next year, their rate of relapse was studied. As shown in Figure 14, patterns of relapse were similar for all three drugs. This suggests similar biological factors.

Other points of similarity between cigarette smoking and drug abuse are tolerance and physiological dependence. As noted earlier, tolerance to tobacco and nicotine is well documented. This effect is shared by most drugs of abuse. The issue of physiological dependence is more complex, however. Drugs are often abused when no physiological dependence can be detected. However, many drugs are associated with discomfort and a heightened desire ("craving") to take the drug during abstinence. Cigarettes are no exception. Indeed many people will "walk a mile," as the ad says, to get a cigarette when they have exhausted their supply.

Nicotine Is an Abusable Drug

As shown in Chapter 6 there was considerable evidence that cigarette smoking was an orderly behavior and a form of drug abuse in which the critical drug was nicotine. However, much of this evidence was circumstantial and not considered sufficient for an agency of the United States government to officially label cigarette smoking as a form of drug abuse. It had to be determined if nicotine itself met the criteria for being defined as a drug of abuse.

Over the years, standardized procedures were developed to discover if new drugs were likely to be abused. Two different kinds of studies were critical. One type is called "the single-dose study," which compares the effects of one dose of the test drug to the effects of single doses of standard drugs of abuse. Morphine, cocaine, alcohol, and pento-

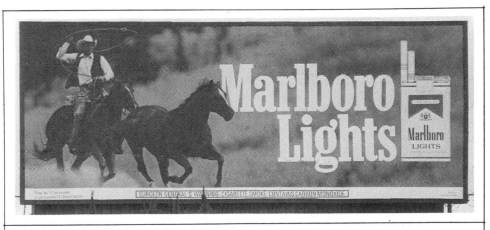

A billboard advertisement for Marlboro Lights carries the mandatory U.S. Surgeon General's warning that cigarette smoke contains carbon monoxide. International tobacco companies currently spend about $2.5 billion a year on advertising, an eightfold increase in the past two decades.

barbital are all highly abused drugs that served as good standards. The other study is called the "self-administration study." This makes the drug available to both animal and human subjects to determine if they will take it voluntarily (i.e., self-administer it).

The single-dose studies compared nicotine in cigarettes to nicotine given intravenously. Volunteers were given research cigarettes to smoke on some test days and were given intravenous nicotine injections on other test days. The subjects were never told what nicotine dose level they were given, or even whether the injection or cigarette contained any nicotine at all. Physiological and psychological measures were taken before and after injections or cigarettes were given.

Several critical findings were made in this study. First, nicotine produced similar effects whether given intravenously or in the form of cigarette smoke. This proved that nicotine is responsible for many of the physiological and psychological effects of tobacco. Second, nicotine produced effects in the brain that allowed volunteers to accurately report when they had been given nicotine and when they had been given a "blank" (placebo). They were also able to rate accurately the strength of the dose. This showed that nicotine was psychoactive. Third, nicotine produced psychological effects of euphoria that were similar to those produced by the standard drugs of abuse. This indicated that nicotine itself was an abusable drug that could produce compulsive behavior. Fourth, subjects with a wide range of experience with drugs of abuse reported that certain effects of nicotine were similar to the effects of cocaine or amphetamine.

Figure 15 illustrates one of the key findings—a psychological effect shared by nicotine and known drugs of abuse. The figure shows a comparison of nicotine to several other drugs of abuse and one drug which is not abused (zomepirac) on a "drug-liking scale." As the figure shows, volunteers can feel the drugs, they like the feelings, and the feelings are stronger with bigger doses. This psychological measure is a hallmark of abusable drugs.

To discover if the psychological effects of nicotine had to do with social and critical factors, certain studies with animals were done. Certain animals are useful for testing

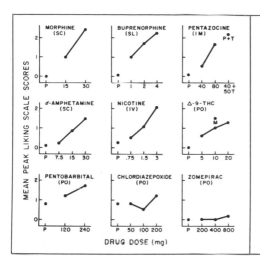

Figure 15. *Average scores on a drug-liking scale are shown for volunteers tested with various drugs. SC indicates a subcutaneous injection. SL indicates a few drops placed beneath the tongue. IM indicates an intramuscular injection. IV indicates an intravenous injection. PO indicates that the drug was swallowed in capsule form. The marijuana (M on the THC graph) was smoked. Note the extraordinarily small dose of nicotine required to register a strong liking response.*

abusable drugs since they usually respond to the drugs the way humans do.

With animals, however, it is generally assumed that the responses are due to the biological effects of the drugs and not to sociocultural factors such as advertising or peer pressure. White rats were trained to press one lever when they received a stimulant and another lever when they received a sedative. When the animals were given nicotine, they pressed the stimulant lever. The larger the dose, the more they pressed the lever. Thus, like humans, the animals were reporting that they could "feel" the injections, that the injections "felt like stimulants," and that they could accurately "rate" the size of the dose.

In the "self-administration studies" human volunteers and animal subjects were permitted to take nicotine intravenously using automatic injection systems. These studies also helped determine whether the drug is important by itself, or if various social and other factors are necessary for the drug to be taken. Both the human volunteers and animals voluntarily took nicotine, and nicotine was found to function as a reinforcer (reward). Figure 16 shows the patterns of nicotine self-administration by human volunteers. The patterns are interesting since they are similar to those found when people smoke cigarettes.

The self-administration studies were important in showing that nicotine could serve as a reinforcer without the other factors that accompany cigarette smoking (taste, oral

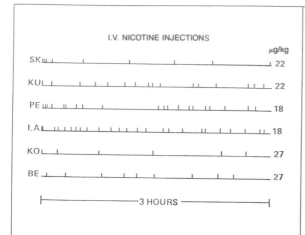

Figure 16. *Patterns of nicotine injections received by six volunteers. Each dose was equal to one typical American cigarette. Actual doses in micrograms per kilogram are shown. Subjects were free to take nicotine by pressing on a lever that delivered a dose to a forearm vein. The patterns of intravenous self-administration of nicotine were very similar to those observed in people smoking cigarettes.*

satisfaction, and peer approval). The results of studies on both human and animal subjects indicate that the biological effects of nicotine are sufficient for it to serve as a reinforcer. In other words, factors unique to humans—television advertising, peer pressure, attempts to prove masculinity or independence—probably have little to do with why nicotine is a reinforcer.

Such studies do not mean that sociocultural factors are not important. They clearly are. Rather, the studies show that nicotine's biological effects make it an ideal drug to be abused and to cause addiction, particularly when it is so widely available.

The results of these studies led the director of the National Institute on Drug Abuse (Dr. William Pollin) to testify to both the United States Senate and the House of Representatives that cigarette smoking is a form of drug abuse in which persons become dependent on the drug nicotine. In its Triennial Report to the United States Congress, the National Institute on Drug Abuse concluded the following:

> In answer to the question, "what are the mechanisms that underlie the compulsive use of tobacco?" the National Institute on Drug Abuse is in agreement with other organizations (e.g., the American Psychiatric Association and the World Health

Organization) that tobacco use can be an addictive form of behavior. In addition to this conclusion, an appraisal of data collected by NIDA's intramural [Addiction Research Center] and extramural [e.g., Johns Hopkins and Harvard] research programs indicates that the behavior is a form of drug abuse in which nicotine is critical. Specifically, it is evident that the role of nicotine in cigarette smoking is similar to the role of cocaine in coca leaf use, of THC in marijuana smoking, and of ethanol in alcoholic beverage consumption.

The results of these studies and of the policy statements by the National Institute on Drug Abuse and the United States Public Health Service (which adopted NIDA's policy) will take time to implement fully. However, changes of medical and political importance already have occurred. New bills were passed in Congress that strengthen warnings on cigarette packages, and treatments for cigarette smoking are being developed based on drug-abuse models.

Major health organizations seeking legislation on smoking and public policy held a news conference in May 1984 to take up the challenge for a "smoke-free society by the year 2000."

A scientist studying the effects of marijuana smoke. Recent studies indicate that smoking "pot" can, like smoking cigarettes, cause lung damage.

A smoker puffs away in a designated smoking area inside a cafeteria at the Pentagon, near Washington, D.C. During the 1980s, concern about the effects on nonsmokers of "passive smoking"—inhaling tobacco fumes secondhand—led to restrictions on smoking in many indoor areas, including restaurants, public buildings, and transportation facilities.

CHAPTER 8

TOBACCO AND THE LAW

*T*he use of tobacco has never been eliminated from a country or major culture into which it was introduced. This is noteworthy since efforts to control the use of tobacco were equal to those used to control other forms of behavior such as crime and religious sin. Opposition to tobacco use took a variety of forms. Surprisingly, severe and immediate punishment proved no more effective in halting smoking than current fears of long-term health consequences.

Rulers came and went, but tobacco remained. Later, governments underwent a conversion of sorts, prompted primarily by the realization that tobacco was an excellent source of revenue—derived either from customs dues (such as those introduced by Cardinal Richelieu in France in 1629) or from the sale of monopolies to deal in tobacco goods. Bohemia was fortified in 1668 with money derived from the tobacco trade, and the Emperor Leopold of Austria used tobacco revenue to finance elaborate hunting expeditions. The large amounts of revenue produced by the manufacture of tobacco products are believed by many to be the main reason why many governments do not fully support anti-tobacco efforts.

Prohibiting Smoking

In the early 1800s Berlin came to regard tobacco use as dangerous and contrary to the character of an orderly and civilized city. As a result, smoking was strictly forbidden, and anyone breaking the law could be arrested and punished by fine, imprisonment, or physical punishment. The prohibition was lifted briefly during a cholera outbreak in the 1830s but was restored and remained in force "because nonsmokers have a clear right not to be annoyed" until 1848. The reference to the rights of nonsmokers is echoed in today's campaign to have greater restrictions placed on public areas in which smoking may take place.

Pope Urban VIII issued a formal decree against tobacco in 1642 and Pope Innocent X issued another in 1650, but clergy as well as laymen continued to smoke. Bavaria prohibited tobacco in 1652, Saxony in 1653, Zurich in 1667, and so on across Europe. The states, like the Church, proved powerless, however, to stem the drug's use.

In Constantinople in 1633 the Sultan Murad IV decreed the death penalty for smoking tobacco. Wherever the Sultan went on his travels or military expeditions, his stopping places were frequently marked by executions of tobacco smokers. Even on the battlefield he was fond of surprising men in the act of smoking: he would punish his own soldiers by beheading, hanging, quartering, or crushing their hands and feet and leaving them helpless between the lines. In spite of the horrors and insane cruelties inflicted by the Sultan, whose blood-lust seemed to increase with age, the passion for smoking persisted in his domain.

The first of the Romanov czars, Mikhail Feodorovich,

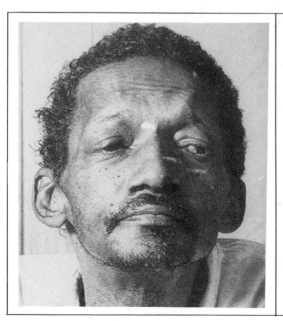

In September 1990 a Mississippi jury decided that cigarette smoking caused the lung-cancer death of Nathan H. Horton in 1987 at the age of 51. The jury, however, refused to award damages to the family of Mr. Horton, saying that they believed both the American Tobacco Company, manufacturer of Pall Malls, the brand Mr. Horton smoked for 26 years, and Mr. Horton himself were responsible for his death. The family had sued the American Tobacco Company for $17 million. In late 1990 at least 55 lawsuits related to smoking deaths were pending against tobacco companies.

also prohibited smoking, under dire penalties, in 1634. "Offenders are usually sentenced to slitting of the nostrils, beatings, or whippings," a visitor to Moscow noted. Yet, in 1698 smokers in Moscow would pay far more for tobacco than English smokers, "and if they lack money, they will sell their clothes for it, to the very shirt."

By 1603 the use of tobacco was well established in Japan and an edict prohibiting smoking was pronounced. As no notice was taken of the edict, still severer measures were taken in 1607 and 1609, by which the cultivation of tobacco was made a criminal offense. Finally, in 1612 it was decreed that the property of any man detected selling tobacco should be handed over to his accuser, and anyone arresting a man conveying tobacco on a packhorse might take both

"Smokey the Bear" burns cigarettes at a "Great American Smokeout" rally on the grounds of Washington Monument. The American Cancer Society sponsored the event to encourage smokers to abandon the habit for the day.

horse and tobacco for his own. Yet in spite of all attempts at repression smoking became so general that in 1615 even the officers in attendance on the Shogun used tobacco. The result was an even sterner measure, to the effect that anyone in the army caught smoking was liable to have his property confiscated.

It was all to no avail. The custom spread rapidly in every direction, until many smokers were to be found even in the Mikado's palace. Finally, even the princes who were responsible for the prohibition took to smoking. Tobacco had won again. In 1625 permission was given to cultivate and plant tobacco. By 1639 tobacco had taken its place as an accompaniment to the ceremonial cup of tea offered to a guest.

By 1725 even the pope was forced to capitulate. Louis Lewin has written on the subject: "Benedict XIII, who himself liked to take snuff, annulled all edicts in order to avoid the scandalous spectacle of dignitaries of the Church hastening out in order to take a few clandestine whiffs in some corner away from spying eyes."

Legislation and Taxation

More recent attempts at control by legislation and taxation have had some documented effects. The average number of cigarettes sold per person, in various countries, is directly affected by price. Scandinavian countries, which have the highest tobacco taxes, have the lowest per capita use of cigarettes. Since much of the price reflects the tax on tobacco, it can be seen that heavy taxation does reduce the use of tobacco. This finding is comforting to economists and psychologists alike since it shows that in some ways tobacco is like other traded goods and addictive drugs: It is an "elastic" commodity. In other words, even though tobacco is used persistently by the population, how heavily it is used varies somewhat with how much it costs.

More recently, "clean air" or "public smoking" legislation has been passed which restricts the use of tobacco from many public places. Most of these laws require certain kinds of places, such as restaurants, to maintain separate smoking and nonsmoking areas for their patrons. While such legislation has been passed to protect nonsmokers from involuntary intake of tobacco smoke, it is also seen by some as a restriction on the freedom of smokers to smoke where and when they please. The proponents of clean air ordinances argue that exposure to tobacco smoke endangers health and well-being. They claim that the right to smoke has neither a moral nor constitutional basis.

The opponents of such legislation argue that the evidence is not great enough to permit such widespread regulation of behavior (a third of adult Americans smoke tobacco). The controversy is probably a healthy one even if smoking is not. Certainly, freedom should be protected and the burden should rest on those who would reduce such freedom. However, as was shown earlier, enough data are in: the evidence that smoking is a specific cause of death and disease is strong, much stronger than the data accepted as sufficient to eliminate lead from gasoline, asbestos from construction materials, and PCB's from water supplies. The danger of being in a room with smokers is now also clear: nonsmokers, in particular children of smokers, are at significantly elevated risk to contract a wide range of diseases.

The availability of tobacco products may also be a factor in their use by increasingly younger males and females. As the U.S. Surgeon General noted in a 1989 report, the number of legal restrictions on children's access to tobacco has decreased over the past twenty-five years. Studies indicate that vendor compliance with minimum-age-of-purchase laws is the exception rather than the rule. More than 90% of smokers today began smoking before the age of 19, and nearly 50% before the age of 15.

As this American Cancer Society poster makes clear, children whose parents smoke are more liable to become smokers themselves than the children of non-smokers. The natural tendency of children to use their parents as role models figures as prominently as peer pressure in the child's decision to begin smoking.

CHAPTER 9

THE TREATMENT OF CIGARETTE SMOKERS

Only about one in four people who receive treatment for cigarette smoking is not smoking one year later. Most people relapse within a few weeks. The success rate for smoking is not much different from that for heroin addiction or alcoholism. Cigarette smoking is a form of drug abuse in which nicotine is the critical drug. As a form of drug abuse, cigarette smoking shares much in common with other forms of drug abuse.

These observations have major implications for how cigarette smokers should be treated. As these implications are considered they may help to improve the efficacy of treatment. The position taken in this volume, that cigarette smoking is a form of drug abuse, is consistent with that of the United States Public Health Service and its various divisions, including the Surgeon General's Office, the Office on Smoking and Health, and the National Institute on Drug Abuse. Other agencies such as the World Health Organization and the American Psychiatric Association hold the same position.

Myths About Cigarette Smokers

Our new knowledge about cigarette smoking should help to dispel some of the myths about why people smoke and why they don't stop smoking. The myths are many and include the following: cigarette smokers are ignorant of the health consequences of smoking; cigarette smokers are less intelligent than nonsmokers; cigarette smokers are psychologically self-destructive; cigarette smokers have poor moral character; cigarette smokers are weak-willed.

In point of fact, cigarette smokers are, by and large, ordinary people whose behavior is, in part, controlled by a powerful environmental factor—nicotine. This is not an excuse for their behavior, but rather a partial explanation for why they may *appear* to some people to be ignorant, irrational, or poorly motivated. Most people who smoke know that something about the cigarette smoke is very important to them, that they will feel genuine discomfort if they don't smoke at certain times, that they may not even be able to perform well at their jobs if they aren't permitted to smoke, and that there are times when they will go to great lengths to be able to smoke. Yet they are given advice by nonsmokers and therapists that include the following: "Why don't you just quit?" "It's just a matter of making up your mind." "Hide your cigarettes." "Chew gum." "Think about something else." "It's harmful so don't do it." "Take up a new hobby to distract yourself." Fortunately, the growing understanding that a powerful drug is involved should help change these views, in smokers and nonsmokers alike.

Admitting that one's behavior is partly controlled by a drug is neither an excuse nor an admission of helplessness. It is a first, critical step in changing the behavior of drug taking. It is hard to treat "weak wills" and "poor moral character," but we can do something about behavior controlled by the drug nicotine.

General Principles of Treatment for Smokers

Since cigarette smoking has been seen as a form of drug abuse, new programs of treatment have been developing rapidly. Many of these apply principles learned over the years in treating drug abuse and alcoholism. One is that the *effect of the drug in the brain is a critical factor* to consider. It is this effect which maintains the compulsive drug-taking behavior of a person for years, or even a lifetime. This is a fact essential to effective treatment programs.

As mentioned earlier, the drug of abuse can be substituted with a similar but safer drug (like nicotine chewing gum), or the effects of the drug can be blocked (for instance, mecamylamine-like drugs can be used as a sort of immunization procedure). Alternatively, if nicotine is providing some therapeutic effects, like stress relief or weight control, then these effects might be replaced by appropriate treatment using either more appropriate drugs or nondrug treatments (for instance, stress management or diet plans).

An equally important principle that has been learned from drug-abuse treatment is that *nondrug factors are also important* and must be considered for effective treatment. It is probably obvious to smokers, if not to others as well, that the pleasures of smoking are many and include taste and smell, the social interactions which they make easier, and even the actual handling of the cigarettes.

Another important set of nondrug factors is the degree to which *tobacco use can become an integral part of behavioral patterns.* For instance, just as a baseball player may find it necessary to tap his bat on home plate, or adjust his cap before feeling comfortable enough to hit, so too might a writer feel unready to face his typewriter without lighting up a cigarette first. The ability to function may be disrupted by the abrupt termination of smoking, and it may take years before the person "feels right" doing many things. Many writers, for instance, feel that to give up smoking would seriously impair their very livelihood. If we consider that a cigarette smoker may light 10,000 cigarettes and take 100,000 puffs per year, it should come as no surprise that smoking may become an important part of the smoker's life.

Treatment of cigarette smoking must address the possibility that *other drugs are also being abused* by the smoker. People who abuse drugs tend to abuse several. For instance, sedatives and alcohol are often abused by the same people, morphine and cocaine are often used jointly, and people who abuse any drug are likely also to smoke cigarettes (more than 90% of opiate and alcohol-dependent persons smoke cigarettes). This finding has led to treatment programs that focus on all of the drugs abused by individuals.

The last major set of findings that have emerged from studies of drug abuse is that quitting can be accompanied by years of *sensitivity and susceptibility to relapse.* In some cases the sensitivity may be partially due to physiological changes. For instance, with the opioid drugs, a secondary and more subtle withdrawal syndrome may persist for six months or more after the initial withdrawal has subsided. Additionally, environmental stimuli may serve as cues which elicit withdrawal-associated discomfort. When this happens, relapse is more likely. In this regard, studies with both opioid drugs and tobacco have shown that stimuli associated with the respective form of drug taking (pictures of needles for the opiate users, smell of cigarette smoke for the cigarette smoker) can elicit such responses: both physiological responses, such as increased heart rate and sweating and psychological responses, such as discomfort and desire to take the drug.

Other kinds of stimuli may also enhance the likelihood of relapse. These include social situations, stress, and anxiety. In the context of drug abuse, relapse is simply one characteristic. It does not mean that treatment has failed, or that the person is a failure. It simply marks a point to restore, possibly with modification, treatment.

If all of the above does not make the treatment of cigarette smoking seem like a hopeful endeavor, at least it should provide an appreciation for why success rates are generally considered good if a particular program achieves 30% abstinence on a one-year follow-up.

Dr. Sharon Hall has developed a program that includes counseling in stress, relapse situations, and chains of behavior. In addition, her subjects are given nicotine-delivering chewing gum as a substitute for cigarettes. Smokers are also asked to smoke cigarettes rapidly to the point of nausea. Dr. Hall found that the combination of all three treatments was most effective. Her study also showed that recognition of the complexity of smoking behavior makes treatment easier. Interestingly, some of Dr. Hall's insights and strategies evolved from her earlier work treating opioid-dependent persons.

Nicotine-Delivering Chewing Gum: Some Facts

In early 1984 the Food and Drug Administration (FDA) approved a nicotine-delivering chewing gum for use in the treatment of cigarette smoking. The gum, developed by a Swedish company, had been under study for more than a decade. It was approved for use in several European countries and Canada several years before its approval in the United States, which is generally more cautious than other countries in approving new drugs.

Chewing the nonnutritive gum, called Nicorette, releases nicotine, and the amount of the drug released is directly related to the rate and intensity of the chewing. The maximum amount of nicotine that may be extracted from one piece is two milligrams, approximately the amount obtained from a cigarette. The gum is buffered so that the nicotine is well absorbed through the buccal mucosae (skin in the mouth and under the tongue). Its effects can be felt within a few minutes of active chewing. While other products have been advertised as substitutes for cigarettes, this is the only one that actually delivers nicotine.

Many smokers say that they would like to quit if they could do so without excessive discomfort. The gum was a major medical advance in giving people the freedom of that choice. Many studies have shown that the gum significantly reduces the discomfort and desire to smoke that accompany quitting. Additionally, the gum may replace some of the pleasure and benefit derived from smoking.

Just as simply giving methadone to opioid abusers without consideration of their particular problem and without supplementary treatment is not terribly effective, we can expect that similar use of the nicotine gum would produce disappointing results. The first and most obvious consideration is that the gum will probably be of little use to people who smoke only a few cigarettes. In their case the nicotine itself is relatively unimportant. Here, a simple test such as the Fagerstrom tolerance questionnaire (see page 113) appears to be a useful predictor.

It is likely that persons who score high on the questionnaire (in other words, are highly dependent on nicotine itself) will be better candidates for effective treatment by the gum. There are, however, a vast array of nonnicotine factors that should be considered. For instance, it may do little good to alleviate a few days of discomfort if the person gains 30 pounds or becomes unable to perform effectively at work. Such people are likely to relapse. The gum should be given in conjunction with a complete program for the treatment of smoking, perhaps tied to some of the more conventional treatments.

However, the gum should be used with the same caution as other medicines. It contains nicotine, and as we have seen, nicotine is a powerful psychoactive drug. While it is in a form that appears safer and more manageable than the nicotine delivered in conjunction with tar, CO, cyanide, and so forth, nicotine is nonetheless a drug with potential toxic effects. Persons with cardiovascular problems, and women who are pregnant, in particular, should be aware that chewing the gum involves some of the same health risks as smoking cigarettes.

The gum should not be viewed as a long-term substitute for most people. In fact, there is the proven possibility of persons becoming dependent on the gum itself. While it can be argued that these people are probably better off than they were while smoking, clearly they are still exposing themselves to an environmental toxin.

Safer Cigarettes: Are They?

Even among cigarette smokers, it has become fashionable to be "health conscious." That is, to eat right, to drink right, to exercise, and even to smoke right. As a result cigarette advertisers began to associate tobacco products with fresh air, mountain streams, sports, dance, and exercise. People, in turn, have grown more sophisticated regarding health care. They count calories, milligrams of sodium in their diets, grams of protein, and perhaps even keep track of the air-quality indexes.

One way that people have grown sophisticated is that they understand that in most cases the critical factor is the *amount* of the substance which they ingest. They understand that everyone is exposed to some cancer-causing agents, for instance, but that the likelihood of getting cancer is related to the amount of those agents. Similarly, many cigarette smokers know that the hazards of smoking are related to the amount of tobacco-containing toxins that they ingest.

Unfortunately, this sophistication in the self-monitoring of what we eat, drink, and smoke has made cigarette smokers easy targets for the "safer cigarette" ad campaigns. Ironically, the federal government, in an effort to provide health information through the Federal Trade Commission (FTC), has become an inadvertent partner in this deception. The FTC oversees the testing of cigarettes and publishes estimates of tobacco contents (such as tar and nicotine). In fact, when one cigarette company became concerned that another company was going too far with its misleading claims, the concerned company sought to sue the FTC and not the other company.

The FTC began to provide lists of its estimates of the amount of tar and nicotine delivered by various brands of cigarettes following the first report of the Surgeon General. Consumers began to smoke the brands containing lower tar and nicotine. As a result, in the last few decades tar and nicotine levels of the average cigarettes sold in the United States have dropped nearly 50%.

On the surface, these trends seem quite promising. If

people are unable to quit, at least they are being exposed to smaller amounts of the toxins as the cigarettes have grown steadily "safer." But *have* the cigarettes grown safer? What do the FTC values mean?

The answer to the first question is that cigarettes are probably not safer than they were several decades ago. There is sound evidence that intake of tar and nicotine has little to do with published levels. People are choosing cigarettes based on misleading information. The problem is that the FTC estimates have little to do with either the content of tobacco or the way people smoke cigarettes. Rather, the FTC estimates are based on the amount of smoke obtained by cigarette-smoking machines, which do not change the way they smoke in response to taste and strength factors. Clearly, testing and advertising methods must be changed if consumers are to be truthfully informed of what they are smoking.

Precautions for Smokers Not Ready to Quit

This was the topic of a pamphlet issued by the Addiction Research Foundation of Toronto. The pamphlet made the

Former smoker Larry Hagman, star of the television series Dallas, *holds copies of the video "Stop Smoking for Life," a self-help program which he developed with researchers at the University of California, Los Angeles. During the years prior to 1960 many celebrities appeared in advertisements for tobacco companies, claiming that smoking had contributed to their success. Today's media personalities, however, are more likely to support antismoking campaigns.*

following recommendations to lessen the medical risk of those cigarette smokers who were not yet ready to quit smoking.

1. Smoke as few cigarettes as possible no matter what their yield (studies show that people who smoke 40 to 50 cigarettes a day have been able to successfully cut back to 10 per day). Also avoid smoking more than two cigarettes per hour at any time of the day.

2. Smoke the lowest-yield cigarettes that you find acceptable, realizing that it may take weeks to get used to them. The greater the decrease in yields the better: differences of only 2 milligrams tar and 0.1 milligrams nicotine are too small to be important. (Note that unless step 1 is followed, this will be futile.)

3. Do not block vent holes on filters.

4. Take fewer puffs per cigarette.

5. Leave longer butts (the last part of a cigarette delivers the highest yields).

6. Avoid inhaling; if you do inhale, take more shallow puffs.

7. Do not keep the cigarette in the mouth between puffs.

Additionally, the primary author of the pamphlet, Dr. Lynn Kozlowski, developed a method whereby smokers can check to see if they are inadvertently blocking ventilation holes. The method is simple. Immediately after smoking a cigarette, inspect the butt end of the cigarette. A yellow stain surrounded by a ring of clean white filter material indicates that the vent holes were open and delivering air to the mouth (and thus less smoke per puff). If the stain covers the end of the butt, then the holes were all blocked. If the stain extends to the filter on one side of the cigarette, then just the holes on that side were blocked, and so forth. Dr. Kozlowski also discovered that cigarette smokers can gradually cut back their tar and nicotine yields by monitoring the color of the stain on the end of the filter. Darker stains indicate that more tar and nicotine were extracted and reached the mouth. Lighter stains indicate that less smoke was extracted. Smokers are encouraged to smoke their cigarettes in such a way as to leave lighter stains on the end of the filter.

The Fagerstrom Tolerance Questionnaire

ANSWER   SCORE

_____ _____ **1.** How soon after you wake up do you smoke your first cigarette?

_____ _____ **2.** Do you find it difficult to refrain from smoking in places where it is forbidden, e.g., in church, at the library, etc.?

_____ _____ **3.** Which of all the cigarettes you smoke in a day is the most satisfying one?

_____ _____ **4.** How many cigarettes a day do you smoke?

_____ _____ **5.** Do you smoke more during the morning than during the rest of the day?

_____ _____ **6.** Do you smoke if you are so ill that you are in bed most of the day?

_____ _____ **7.** What brand do you smoke?

_____ _____ **8.** How often do you inhale?

Scoring

The questions are scored so that higher points indicate a higher level of addiction to cigarettes. A score of 6 or more indicates a high level of physical dependence.

1: One point is assigned to smoking within 30 minutes.

2, 5 and 6: Items are scored with one point for yes answers.

3: One point is assigned for answering "the first cigarette in the morning."

4: Smokers are divided as light, 1-15 cigarettes; moderate, 16-25; and heavy, more than 25 (1 to 3 points).

7: The brands are classified into three categories with low, medium, and high nicotine levels (1 to 3 points).

8: Always—2 points; sometimes—1 point; never—0 points.

Source: K. O. Fagerstrom, "Measuring Degree of Physical Dependence on Tobacco," *Addictive Behaviors,* Vol. 3 (1978), p. 235.

Tips for Quitting Smoking

As should now be evident, good cigarette treatment programs require much psychological and medical knowledge. However, there are some basic strategies which remain the core of modern treatment programs. These strategies should be applied by smokers themselves to help them in their personal efforts. Since most people who quit do so without the specific aid of a treatment program, widespread knowledge of these strategies and some of the information provided in this volume could do much to reduce cigarette smoking.

    1.   Discover some of the reasons that you smoke. Discussing your smoking behavior with another person may be helpful. The Fagerstrom tolerance questionnaire (see page 113) will help you to gauge your level of dependence; another questionnaire called Reasons for Smoking is available from the Office on Smoking and Health (see Chapter 10).

    2.   Self-monitor your cigarette smoking behavior for one or two weeks. An index card may be placed inside the wrapper of your cigarette package. Then, each time you smoke a cigarette, you should write the time, place, and reason for smoking.

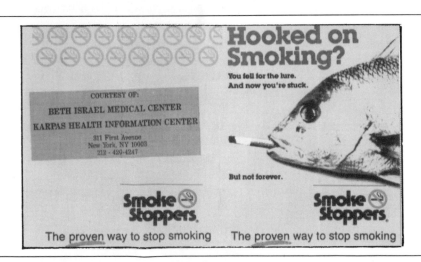

A New York City hospital advertises their sponsorship of Smoke Stoppers, a nationwide self-help program for smokers who want to quit. Seeking help for a recognized addiction is the first step toward successful recovery.

3. Steps one and two will provide you with important information about your own smoking pattern. The next step is to cut back your cigarette intake to about 10 cigarettes per day. Make sure that all smoking that you do is according to *your* prearranged plan (for instance, only smoking in certain places and at certain times). If you don't plan, but simply "hold out as long as possible" between each cigarette, each cigarette will become of exaggerated importance and benefit.

4. After a few weeks of learning to control your smoking behavior and smoking at reduced levels, quit—cold turkey. Studies show that gradual reduction may result in more persistent craving, even a year later. Physical discomfort from abrupt quitting following a one-half-pack-a-day pattern should be minimal. Try to avoid quitting at a particularly stressful time or when you may be attending a party in which cigarettes and alcohol will be available. Situations such as these may hinder your efforts.

5. For the first few days after quitting, treat yourself well. You have done one of the most important things you may ever do for yourself; you should take pride in it and reward yourself. Even if you are prone to gaining weight, this may be an appropriate time to reward your quitting smoking with good (though not a lot of) food.

6. Seek the aid of friends and family. The biggest single correlate of both starting and quitting smoking is peers who are doing the same. Quitting with another person is a particularly useful approach. Otherwise, at least enlisting the social support of others before quitting can be very beneficial.

7. Most people who relapse do so within the first three months of quitting. During this period, in particular, try to avoid temptations (smelling a friend's tobacco smoke) and relapse situations (parties and stressful situations). Remember too that drugs used previously in conjunction with smoking may promote relapse.

8. Finally, and perhaps most importantly, if you relapse, recognize that you have not failed. Relapse is simply one normal element of breaking a pattern of behavior in which an abusable drug is involved. The important thing is to resume your own efforts as soon as possible. Now you will have some experience and may design a better treatment for yourself in light of your new knowledge.

AP/WIDE WORLD PHOTOS

A group of Boston first-graders participating in the Smoke-Free Class of 2000 project are honored at a special ceremony in May 1989. Sponsored by the American Cancer Society, the American Heart Association, and the American Lung Association, the project aims to keep these children from smoking until they graduate from high school.

CHAPTER 10

FURTHER HELP ON QUITTING SMOKING

*T*his chapter contains descriptions of the major national smoking cessation and prevention programs in the United States. Federally funded treatment and information programs are also described. This chapter has been adopted from one jointly published by the Office on Smoking and Health; the National Cancer Institute; the National Heart, Lung, and Blood Institute; and the Centers for Disease Control.

American Cancer Society

Every year, the American Cancer Society's 58 divisions and 3,100 local units reach millions of adults and young people with smoking education, prevention, and cessation programs. In addition, millions of pamphlets, posters, and exhibits on smoking are distributed each year. Smokers are targeted in November by the Great American Smokeout, in April by the annual Cancer Crusade, and throughout the year by mass media efforts and physician counseling.

Educational programs and smoking cessation clinics are conducted on the local level in the context of a broader comprehensive smoking control program that may also include, among other things, mass media efforts, legislative initiatives, professional education, response/referral systems, and/or a smokers' telephone "quit-line." Heavy emphasis is placed on preparing others (e.g., hospitals or industry) to assume primary responsibility for helping smokers to quit.

Education in the prevention of smoking is carried out principally in the schools, because young people presumably have not yet established the smoking habit. Programs begin in the elementary schools, where students are just learning about their bodies and good health habits, and continue through high school. Among the more widely disseminated smoking prevention programs developed by the American Cancer Society (ACS) are:

An Early Start to Good Health, a series of teaching units designed to promote positive health behavior in grades K–3.

ACS Health Network, a followup series to *An Early Start to Good Health*, designed for students in grades 4–6.

Cigarette Smoking—Take It or Leave It, a filmstrip with accompanying cassette, transparencies, and teacher's guide, focusing on values clarification and decision-making.

Who's in Charge Here?, a film and teacher's guide illustrating the immediate physiological effects of smoking and refuting common misconceptions about the associated health hazards.

The ACS also has packaged and disseminated two smoking cessation programs, one for adults and one for teens.

Smoking Cessation Clinics for High School Students

In response to numerous requests from school administrators for a smoking cessation program for high school students, in 1976 the American Cancer Society and the Iowa Department of Education adapted the ACS adult cessation clinic for use by high school students (grades 9–12). Emphasis in this program is on teaching young people decision-making, values clarification, and other self-management skills to assist them in quitting smoking, and on coordinating group support activities to reinforce nonsmoking behavior.

American Lung Association

American Lung Association (ALA) activities are designed to clean up the air in indoor and outdoor environments, to discourage smoking, and to prevent lung disease caused by microorganisms. Lung Association staff conduct public education programs, seminars, and workshops for health professionals; answer public inquiries; finance research; and work with other organizations to clean up the atmosphere.

Most local units of the ALA have a program specialist who provides consultant services and materials to schools for developing smoking education curricula and for implementing biofeedback and peer-teaching programs about smoking. Other activities include sponsoring legislation in support of nonsmokers' rights, and providing help for smokers interested in quitting. Books, pamphlets, puzzles, posters, buttons, bookmarks, desk signs, and films are commonly available through local units.

Freedom from Smoking

In January 1981 the American Lung Association launched a new self-help program to help smokers kick the habit. Freedom from Smoking is based on two colorful, extensively illustrated manuals: "Freedom from Smoking in 20 Days," a nuts-and-bolts, day-by-day approach to quitting, and "A Lifetime of Freedom from Smoking," which helps smokers reinforce and maintain their new nonsmoking lifestyles. In the second manual, tensions and events that may cause backsliding are anticipated, and counter-strategies are spelled out. The manuals emphasize making smoking cessation a life-enhancing process that can be satisfying and fun. Ways to change and improve eating habits, reduce stress, assert feelings, and savor life are highlighted as elements of a healthier lifestyle. Freedom from Smoking also places heavy emphasis on the maintenance of nonsmoking.

Biofeedback

The week-long Biofeedback Smoking Education program was developed and piloted during the 1977–78 school year by the New Hampshire Lung Association. Since then, the program or modified forms of it have been adopted by Lung Association units across the country.

The program, for students in grades 7–12, emphasizes the immediate physiological effects of smoking. The objective is to stimulate a positive change in students' smoking behavior and attitudes and to reinforce the behavior and attitudes for nonsmoking students.

Student Teach Student

The Student Teach Student program was originally developed by the Wisconsin Lung Association and continues to be used throughout the state of Wisconsin and in other localities. The program features high school volunteers who make presentations to fifth and sixth graders about the pros and cons of smoking. Nonsmoking high school students are chosen by a faculty member and are trained after school for five one-hour periods by high school teachers who volunteer their time. Training consists of discussions of the health effects of smoking, how peer pressure influences youth to smoke, the history of tobacco, and how to answer grade schoolers' questions.

Seventh-Day Adventist Church

Five-Day Plan

In 1959 a Seventh-Day Adventist physician and a minister teamed their efforts to present a program on the harmful effects of smoking and the psychological and emotional aspects of the tobacco habit. The result was the Five-Day Plan, a series of five consecutive 90-minute sessions designed to show smokers how to beat the habit in all four dimensions of life—physical, mental, social, and spiritual. The principle is one of discovering how to use natural methods to quit smoking without the use of drugs. The program is usually presented by a team consisting of a Seventh-Day Adventist physician and a clergyman. It is often conducted in cooperation with, or cosponsored by, other civic or church groups.

Talks, films, group discussion therapy, a special day-to-day pocket manual, and other aids are all part of the format of this program. Participants are urged to quit "cold turkey" and to make a strong, forceful decision not to relapse. Among the techniques advocated during the withdrawal period are: frequent warm baths; increased fluid intake; regular eating and sleeping habits; extra exercise, particularly after meals; avoidance of all sedatives and stimulants, including alcohol and caffeine; dietary restrictions; extra vitamins; and prayer.

For further information contact:

General Conference of Seventh-Day Adventists
Health and Temperance Department
6840 Eastern Avenue, N.W.
Washington, D.C. 20012

U.S. Department of Health and Human Services Centers for Disease Control

The Centers for Disease Control (CDC) assists public and private agencies to prevent and reduce problems of public health significance. In its public health effort to combat the current leading causes of death and disability, CDC's Center for Health Promotion and Education has supported Health Education Risk Reduction activities to reduce cigarette smoking, a leading risk factor for many of today's health problems.

Primary Grades Health Curriculum Project

The Center for Health Promotion and Education and the American Lung Association cooperated in the development of the Primary Grades Health Curriculum Project (the Seattle Project) for children in kindergarten through third grade. Although smoking is not the primary focus, the subject of smoking and health is integrated into this comprehensive curriculum, designed to teach good health practices through an understanding of and appreciation for body systems and functions and the development of a positive self-image.

School Health Curriculum Project

Originally developed and pilot-tested by the National Clearinghouse for Smoking and Health (now the Office on Smoking and Health) in 1969 in San Ramon, California, the School Health Curriculum Project (SHCP), also known as the Berkeley Project, is now funded by a wide range of organizations and individuals.

The curriculum consists of four units of study for grades 4 through 7, each of which is organized around a particular body system and can be taught in an 8- to 10-week period. The four units, "Our Decisions, Nutrition, Our Health," "About Our Lungs and Our Health," "Our Health and Our Hearts," and "Living Well with Our Nervous Systems," focus respectively on the digestive, respiratory, cardiovascular, and nervous systems. Like the Primary Grades Health Curriculum Project, which was designed to be used in coordination with this program, the School Health Curriculum Project integrates smoking education into all four units. Each unit covers the physiology of the body system being studied, how the body system can be affected by man's abuse of the environment, how it is possible to abuse the body by individual actions like smoking, and how to take care of the body for maximum health. All units are specifically correlated with other subjects in the curriculum, such as art, music, mathematics, social studies, and language skills.

For further information contact:

Center for Health Promotion and Education
Centers for Disease Control
Building 14
1600 Clinton Road, N.E.
Atlanta, GA 30333

Other Federal Resources in Smoking and Health

Several agencies within the United States Public Health Service offer information and referral services on smoking and health as part of their mission to reduce the incidence of smoking-related diseases and conditions. In addition to the programs sponsored by the Centers for Disease Control, described previously, the following federal agencies offer a wide range of services to the general public as well as to health professionals.

National Cancer Institute

The Cancer Information Service (CIS) is a toll-free telephone inquiry system that supplies information about cancer and cancer-related resources to the general public and to cancer patients and their families. CIS offices are in locations associated with major cancer centers across the United States. In addition to providing assistance over the telephone, each CIS office offers free printed materials on subjects ranging from types of cancer and treatment to advice on how to talk with cancer patients. Some materials are available in Spanish and other languages.

For answers to questions about smoking cessation, referrals to local services, or general information on smoking, you can call the Cancer Information Service by dialing the number for your state:

Alabama: 1-800-292-6201	Illinois: 1-800-972-0586
Alaska: 1-800-638-6070	Kentucky: 1-800-432-9321
California: 1-800-252-9066	Maine: 1-800-225-7034
(from area codes 213, 714	Maryland: 1-800-492-1444
and 805 only)	Massachusetts: 1-800-952-7420
Colorado: 1-800-332-1850	Michigan: 1-800-482-4959
Connecticut: 1-800-922-0824	Minnesota: 1-800-582-5262
Delaware: 1-800-523-3586	New Hampshire: 1-800-225-8034
District of Columbia: 636-5700	New Jersey: northern: 1-800-223-1000
Florida: 1-800-432-5953	southern: 1-800-523-3586
Georgia: 1-800-327-7332	New York State: 1-800-462-7255
Hawaii: Oahu 524-1234	New York City: (212) 794-7982
Neighbor Islands: call collect	North Carolina: 1-800-672-0943

North Dakota: 1-800-328-5188
Ohio: 1-800-282-6522
Pennsylvania: 1-800-822-3963
South Dakota: 1-800-328-5188

Texas: 1-800-392-2040
Vermont: 1-800-225-7034
Washington: 1-800-552-7212
Wisconsin: 1-800-362-8038

FOR ALL OTHER STATES: 1-800-638-6694

National Heart, Lung, and Blood Institute

For information on heart disease and smoking contact:

Public Inquiries and Reports Branch
National Heart, Lung, and Blood Institute
Bldg. 31, Room 4A-21
National Institutes of Health
Bethesda, MD 20205
(301) 496-4236

Office on Smoking and Health

The Office on Smoking and Health (OSH) is the principal agency within the Department of Health and Human Services concerned with the problems of smoking, tobacco, and tobacco use and its effect on health. One of the primary functions of the office is its Health Information Program. Consisting of two distinct entities, one for public information and the other for scientific and technical information, it develops and disseminates a wide variety of information resources for specific audiences.

The Public Information Program develops print and nonprint materials for the general public; the Technical Information Program is concerned with the collection and dissemination of scientific data for researchers and other health professionals.

The office receives and responds to requests for information on all aspects of smoking and health. Inquiries can be referred to the office at the following addresses:

Public Inquiries
Office on Smoking and Health
Park Bldg., Room 1-58
5600 Fishers Lane
Rockville, MD 20857
(301) 443-1575

Technical Inquiries
Technical Information Center
Office on Smoking and Health
Park Bldg., Room 1-16
5600 Fishers Lane
Rockville, MD 20857
(301) 443-1690

Appendix I

POPULATION ESTIMATES OF LIFETIME AND CURRENT NONMEDICAL DRUG USE, 1988

	12-17 years (pop. 20,250,000)				18-25 years (pop. 29,688,000)			
	%	Ever Used	%	Current User	%	Ever Used	%	Current User
Marijuana & Hashish	17	3,516,000	6	1,296,000	56	16,741,000	16	4,594,000
Hallucinogens	3	704,000	1	168,000	14	4,093,000	2	569,000
Inhalants	9	1,774,000	2	410,000	12	3,707,000	2	514,000
Cocaine	3	683,000	1	225,000	20	5,858,000	5	1,323,000
Crack	1	188,000	+	+	3	1,000,000	1	249,000
Heroin	1	118,000	+	+	+	+	+	+
Stimulants*	4	852,000	1	245,000	1	3,366,000	2	718,000
Sedatives	2	475,000	1	1 23,000	6	1,633,000	1	265,000
Tranquilizers	2	413,000	+	+	8	2,319,000	1	307,000
Analgesics	4	840,000	1	182,000	9	2,798,000	1	440,000
Alcohol	50	10,161,000	25	5,097,000	90	26,807,000	65	19,392,000
Cigarettes	42	8,564,000	12	2,389,000	75	22,251,000	35	10,447,000
Smokeless Tobacco	15	3,021,000	4	722,000	24	6,971,000	6	1,855,000

* Amphetamines and related substances
+ Amounts of less than .5% are not listed
  Terms: Ever Used: used at least once in a person's lifetime.
          Current User: used at least once in the 30 days prior to the survey.

Source: National Institute on Drug Abuse, August 1989

26+ years (pop. 148,409,000)				TOTAL (pop. 198,347,000)			
%	Ever Used	%	Current User	%	Ever Used	%	Current User
31	45,491,000	4	5,727,000	33	65,748,000	6	11,616,000
7	9,810,000	+	+	7	4,607,000	+	+
4	5,781,000	+	+	6	1,262,000	1	1,223,000
10	14,631,000	1	1,375,000	11	21,171,000	2	2,923,000
+	+	+	+	1	2,483,000	+	484,000
1	1,686,000	+	+	1	1,907,000	+	+
7	9,850,000	1	791,000	7	4,068,000	1	1,755,000
3	4,867,000	+	+	4	6,975,000	+	+
5	6,750,000	1	822,000	5	9,482,000	1	1,174,000
5	6,619,000	+	+	5	10,257,000	1	1,151,000
89	131,530,000	55	81,356,000	85	168,498,000	53	105,845,000
80	118,191,000	30	44,284,000	75	149,005,000	29	57,121,000
13	19,475,000	3	4,497,000	15	29,467,000	4	7,073,000

Appendix II

DRUGS MENTIONED MOST FREQUENTLY BY HOSPITAL EMERGENCY ROOMS, 1988

	Drug name	Number of mentions by emergency rooms	Percent of total number of mentions
1	Cocaine	62,141	38.80
2	Alcohol-in-combination	46,588	29.09
3	Heroin/Morphine	20,599	12.86
4	Marijuana/Hashish	10,722	6.69
5	PCP/PCP Combinations	8,403	5.25
6	Acetaminophen	6,426	4.01
7	Diazepam	6,082	3.80
8	Aspirin	5,544	3.46
9	Ibuprofen	3,878	2.42
10	Alprazolam	3,846	2.40
11	Methamphetamine/Speed	3,030	1.89
12	Acetaminophen W Codeine	2,457	1.53
13	Amitriptyline	1,960	1.22
14	D.T.C. Sleep Aids	1,820	1.14
15	Methadone	1,715	1.07
16	Triazolam	1,640	1.02
17	Diphenhydramine	1,574	0.98
18	D-Propoxyphene	1,563	0.98
19	Hydantoin	1,442	0.90
20	Lorazepam	1,345	0.84
21	LSD	1,317	0.82
22	Amphetamine	1,316	0.82
23	Phenobarbital	1,223	0.76
24	Oxycodone	1,192	0.74
25	Imipramine	1,064	0.66

Source: Drug Abuse Warning Network (DAWN), Annual Data 1988

Appendix III

DRUGS MENTIONED MOST FREQUENTLY BY MEDICAL EXAMINERS (IN AUTOPSY REPORTS), 1988

	Drug name	Number of mentions in autopsy reports	Percent of total number of drug mentions
1	Cocaine	3,308	48.96
2	Alcohol-in-combination	2,596	38.43
3	Heroin/Morphine	2,480	36.71
4	Codeine	689	10.20
5	Diazepam	464	6.87
6	Methadone	447	6.62
7	Amitriptyline	402	5.95
8	Nortriptyline	328	4.85
9	Lidocaine	306	4.53
10	Acetaminophen	293	4.34
11	D-Propoxyphene	271	4.01
12	Marijuana/Hashish	263	3.89
13	Quinine	224	3.32
14	Unspec Benzodiazepine	222	3.29
15	PCP/PCP Combinations	209	3.09
16	Diphenhydramine	192	2.84
17	Phenobarbital	183	2.71
18	Desipramine	177	2.62
19	Methamphetamine/Speed	161	2.38
20	Doxepin	152	2.25
21	Aspirin	138	2.04
22	Imipramine	137	2.03
23	Hydantoin	98	1.45
24	Amphetamine	87	1.29
25	Chlordiazepoxide	76	1.12

Source: Drug Abuse Warning Network (DAWN), <u>Annual Data 1988</u>

Appendix IV

NATIONAL HIGH SCHOOL SENIOR SURVEY, 1975-1989

	High School Senior Survey Trends in Lifetime Prevalence Percent Who Ever Used				
	Class of 1975	Class of 1976	Class of 1977	Class of 1978	Class of 1979
Marijuana/Hashish	47.3	52.8	56.4	59.2	60.4
Inhalants	NA	10.3	11.1	12.0	12.7
Inhalants Adjusted	NA	NA	NA	NA	18.2
Amyl & Butyl Nitrites	NA	NA	NA	NA	11.1
Hallucinogens	16.3	15.1	13.9	14.3	14.1
Hallucinogens Adjusted	NA	NA	NA	NA	17.7
LSD	11.3	11.0	9.8	9.7	9.5
PCP	NA	NA	NA	NA	12.8
Cocaine	9.0	9.7	10.8	12.9	15.4
Crack	NA	NA	NA	NA	NA
Other cocaine	NA	NA	NA	NA	NA
Heroin	2.2	1.8	1.8	1.6	1.1
Other Opiates*	9.0	9.6	10.3	9.9	10.1
Stimulants*	22.3	22.6	23.0	22.9	24.2
Stimulants Adjusted*	NA	NA	NA	NA	NA
Sedatives*	18.2	17.7	17.4	16.0	14.6
Barbiturates*	16.9	16.2	15.6	13.7	11.8
Methaqualone*	8.1	7.8	8.5	7.9	8.3
Tranquilizers*	17.0	16.8	18.0	17.0	16.3
Alcohol	90.4	91.9	92.5	93.1	93.0
Cigarettes	73.6	75.4	75.7	75.3	74.0

Stimulants adjusted to exclude inappropriate reporting of nonprescription stimulants; stimulants = amphetamines and amphetamine-like substances.
*Only use not under a doctor's orders included.

Source: National Institute on Drug Abuse, National High School Senior Survey: "Monitoring the Future," 1989

NATIONAL HIGH SCHOOL SENIOR SURVEY, 1975-1989

High School Senior Survey
Trends in Lifetime Prevalence
Percent Who Ever Used

Class of 1980	Class of 1981	Class of 1982	Class of 1983	Class of 1984	Class of 1985	Class of 1986	Class of 1987	Class of 1988	Class of 1989
60.3	59.5	58.7	57.0	54.9	54.2	50.9	50.2	47.2	43.7
11.9	12.3	12.8	13.6	14.4	15.4	15.9	17.0	16.7	17.6
17.3	17.2	17.7	18.2	18.0	18.1	20.1	18.6	17.5	18.6
11.1	10.1	9.8	8.4	8.1	7.9	8.6	4.7	3.2	3.3
13.3	13.3	12.5	11.9	10.7	10.3	9.7	10.3	8.9	9.4
15.6	15.3	14.3	13.6	12.3	12.1	11.9	10.6	9.2	9.9
9.3	9.8	9.6	8.9	8.0	7.5	7.2	8.4	7.7	8.3
9.6	7.8	6.0	5.6	5.0	4.9	4.8	3.0	2.9	3.9
15.7	16.5	16.0	16.2	16.1	17.3	16.9	15.2	12.1	10.3
NA	NA	NA	NA	NA	NA	NA	5.4	4.8	4.7
NA	NA	NA	NA	NA	NA	NA	14.0	12.1	8.5
1.1	1.1	1.2	1.2	1.3	1.2	1.1	1.2	1.1	1.3
9.8	10.1	9.6	9.4	9.7	10.2	9.0	9.2	8.6	8.3
26.4	32.2	35.6	35.4	NA	NA	NA	NA	NA	NA
NA	NA	27.9	26.9	27.9	26.2	23.4	21.6	19.8	19.1
14.9	16.0	15.2	14.4	13.3	11.8	10.4	8.7	7.8	7.4
11.0	11.3	10.3	9.9	9.9	9.2	8.4	7.4	6.7	6.5
9.5	10.6	10.7	10.1	8.3	6.7	5.2	4.0	3.3	2.7
15.2	14.7	14.0	13.3	12.4	11.9	10.9	10.9	9.4	7.6
93.2	92.6	92.8	92.6	92.6	92.2	91.3	92.2	92.0	90.7
71.0	71.0	70.1	70.6	69.7	68.8	67.6	67.2	66.4	65.7

Appendix V

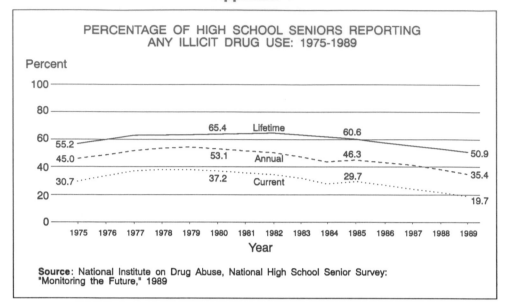

PERCENTAGE OF HIGH SCHOOL SENIORS REPORTING
ANY ILLICIT DRUG USE: 1975-1989

Percent

Source: National Institute on Drug Abuse, National High School Senior Survey:
"Monitoring the Future," 1989

Appendix VI

DRUG ABUSE AND AIDS

An estimated 25 percent of all cases of acquired immunodeficiency syndrome, or AIDS, are intravenous (IV) drug abusers. This group is the second largest at risk for AIDS, exceeded only by homosexual, and bisexual men. And the numbers may be growing. Data for the first half of 1988 show that IV drug abusers made up about 31 percent of the total reported cases.

"... the number of IV drug users with AIDS is doubling every 14-16 months."

According to the National Institute on Drug Abuse (NIDA). There are 1.1 to 1.3 million IV drug users in the United States, and, so far, about 17,500 have developed AIDS. Thousands more are infected with the virus that causes this fatal illness, which kills by destroying the body's ability to fight disease.

Currently, the number of IV drug users with AIDS is doubling every 14-16 months. Although the numbers of IV drug users who carry the AIDS virus varies from region to region, in some places the majority may already be infected. In New York City, for example, 60 percent of IV drug users entering treatment programs have the AIDS virus.

Among IV drug abusers, the AIDS virus is spread primarily by needle sharing. As long as IV drug abusers are drug dependent, they are likely to engage in needle sharing. Thus, the key to eliminating needle sharing—and the associated spread of AIDS—is drug abuse treatment to curb drug dependence. NIDA is working to find ways to get more IV users into treatment and to develop new methods to fight drug addiction.

Most non-drug users characteristically associate heroin with IV drug use. However, thousands of others inject cocaine or amphetamines. Recent evidence suggests that IV cocaine use is increasing and that the AIDS virus is spreading in those users. One reason for this may be because cocaine's effects last only a short time. When the drug, which is a stimulant, wears off, users may inject again and again, sharing a needle many times in a few hours. In contrast, heroin users inject once and fall asleep.

". . . IV cocaine use is increasing and the AIDS virus is spreading in those users."

The apparent increase in IV cocaine is especially worrisome, drug abuse experts say, because there are no standard therapies for treating cocaine addiction. Until scientists find effective treatments for this problem, the ability to control the spread of AIDS will be hampered.

TRANSMISSION

Needle Sharing -- Among IV drug users, transmission of AIDS virus most often occurs by sharing needles, syringes, or other "works." Small amounts of contaminated blood left in the equipment can carry the virus from user to user. IV drug abusers who frequent "shooting galleries" — where paraphernalia is passed among several people -- are at especially high risk for AIDS. But, needle sharing of any sort (at parties, for example) can transmit the virus, and NIDA experts note that almost all IV drug users share needles at one time or another.

Because not every IV drug abuser will enter treatment and because some must wait to be treated, IV users in many cities are being taught to flush their "works" with bleach before they inject. Used correctly, bleach can destroy virus left in the equipment.

Sexual Transmission -- IV drug abusers also get AIDS through unprotected sex with someone who is infected. In addition, the AIDS virus can be sexually transmitted from infected IV drug abusers to individuals who do not use drugs. Data from the Centers for Disease Control show that IV drug use is associated with the increased spread of AIDS in the heterosexual population. For example, of all women reported to have AIDS, 49 percent were IV drug users, while another 30 percent -- non-IV drug users themselves -- were sexual partners of IV drug users. Infected women who become pregnant can pass the AIDS virus to their babies. About 70 percent of all children born with AIDS have had a mother or father who shot drugs.

Non-IV Drug Use and AIDS -- Sexual activity has also been reported as the means of AIDS transmission among those who use non-IV drugs (like crack or marijuana). Many people, especially women, addicted to crack (or other substances) go broke supporting their habit and turn to trading sex for drugs. Another link between substance abuse and AIDS is when individuals using alcohol and drugs relax their restraints and caution regarding sexual behavior. People who normally practice "safe" sex may neglect to do so while "under the influence."

Source: U.S. Public Health Service, AIDS Program Office, 1989

Appendix VII

U.S. Drug Schedules*

	Drugs Included	Dispensing Regulations
Schedule I high potential for abuse; no currently accepted medical use in treatment in U.S.; safety not proven for medical use	heroin methaqualone LSD mescaline peyote phencyclidine analogs psilocybin marijuana hashish	research use only
Schedule II high potential for abuse; currently accepted U.S. medical use; abuse may lead to severe psychological or physical dependence	opium morphine methadone barbiturates cocaine amphetamines phencyclidine codeine	written Rx; no refills
Schedule III less potential for abuse than drugs in Schedules I and II; currently accepted U.S. medical use; may lead to moderate or low physical dependence or high psychological dependence	glutethimide selected morphine, opium, and codeine compounds selected depressant sedative compounds selected stimulants for weight control	written or oral Rx; refills allowed
Schedule IV low potential for abuse relative to drugs in Schedule III; currently accepted U.S. medical use; abuse may lead to limited physical dependence or psychological dependence relative to drugs in Schedule III	selected barbiturate and other depressant compounds selected stimulants for weight control	written or oral Rx; refills allowed
Schedule V low potential for abuse relative to drugs in Schedule IV; currently accepted U.S. medical use; abuse may lead to limited physical dependence or psychological dependence relative to drugs in Schedule IV	selected narcotic compounds	OTC/ M.D.'s order

*Established by the U.S. Controlled Substances Act of 1970
Source: U.S. Drug Enforcement Administration

Appendix VIII

Agencies for the Prevention and Treatment of Drug Abuse

UNITED STATES

Alabama
Department of Mental Health
Division of Substance Abuse
200 Interstate Park Drive
P.O. Box 3710
Montgomery, AL 36109
(205) 270-9650

Alaska
Department of Health and
  Social Services
Division of Alcoholism and
  Drug Abuse
P.O. Box H
Juneau, AK 99811-0607
(907) 586-6201

Arizona
Department of Health
  Services
Division of Behavioral Health
  Services
Bureau of Community
  Services
The Office of Substance
  Abuse
2632 East Thomas
Phoenix, AZ 85016
(602) 255-1030

Arkansas
Department of Human
  Services
Division of Alcohol and Drug
  Abuse
400 Donagy Plaza North
P.O. Box 1437
Slot 2400
Little Rock, AR 72203-1437
(501) 682-6656

California
Health and Welfare Agencies
Department of Alcohol and
  Drug Programs
1700 K Street
Sacramento, CA 95814-4037
(916) 445-1943

Colorado
Department of Health
Alcohol and Drug Abuse
  Division
4210 East 11th Avenue
Denver, CO 80220
(303) 331-8201

Connecticut
Alcohol and Drug Abuse
  Commission
999 Asylum Avenue
3rd Floor
Hartford, CT 06105
(203) 566-4145

Delaware
Division of Mental Health
Bureau of Alcoholism and
  Drug Abuse
1901 North Dupont Highway
Newcastle, DE 19720
(302) 421-6101

District of Columbia
Department of Human
  Services
Office of Health Planning and
  Development
1660 L Street NW
Room 715
Washington, DC 20036
(202) 724-5641

Florida
Department of Health and
  Rehabilitative Services
Alcohol, Drug Abuse, and
  Mental Health Office
1317 Winewood Boulevard
Building 6, Room 183
Tallahassee, FL 32399-0700
(904) 488-8304

Georgia
Department of Human
  Resources
Division of Mental Health,
  Mental Retardation, and
  Substance Abuse
Alcohol and Drug Section
878 Peachtree Street
Suite 319
Atlanta, GA 30309-3917
(404) 894-4785

Hawaii
Department of Health
Mental Health Division
Alcohol and Drug Abuse
  Branch
1270 Queen Emma Street
Room 706
Honolulu, HI 96813
(808) 548-4280

Idaho
Department of Health and
  Welfare
Bureau of Preventive
  Medicine
Substance Abuse Section
450 West State
Boise, ID 83720
(208) 334-5934

Illinois
Department of Alcoholism
  and Substance Abuse
Illinois Center
100 West Randolph Street
Suite 5-600
Chicago, IL 60601
(312) 814-3840

Indiana
Department of Mental Health
Division of Addiction Services
117 East Washington Street
Indianapolis, IN 46204-3647
(317) 232-7816

Iowa
Department of Public Health
Division of Substance Abuse
Lucas State Office Building
321 East 12th Street
Des Moines, IA 50319
(515) 281-3641

Kansas
Department of Social
Rehabilitation
Alcohol and Drug Abuse
Services
300 SW Oakley
2nd Floor
Biddle Building
Topeka, KS 66606
(913) 296-3925

Kentucky
Cabinet for Human Resources
Department of Health
Services
Substance Abuse Branch
275 East Main Street
Frankfort, KY 40621
(502) 564-2880

Louisiana
Department of Health and
Hospitals
Office of Human Services
Division of Alcohol and Drug
Abuse
P.O. Box 3868
Baton Rouge, LA 70821-3868
1201 Capital Access Road
Baton Rouge, LA 70802
(504) 342-9354

Maine
Department of Human
Services
Office of Alcoholism and
Drug Abuse Prevention
Bureau of Rehabilitation
5 Anthony Avenue
State House, Station 11
Augusta, ME 04433
(207) 289-2781

Maryland
Alcohol and Drug Abuse
Administration
201 West Preston Street

4th Floor
Baltimore, MD 21201
(301) 225-6910

Massachusetts
Department of Public Health
Division of Substance Abuse
150 Tremont Street
Boston, MA 02111
(617) 727-1960

Michigan
Department of Public Health
Office of Substance Abuse
Services
2150 Apollo Drive
P.O. Box 30206
Lansing, MI 48909
(517) 335-8810

Minnesota
Department of Human
Services
Chemical Dependency
Division
444 Lafayette Road
St. Paul, MN 55155
(612) 296-4614

Mississippi
Department of Mental
Health
Division of Alcohol and Drug
Abuse
1101 Robert E. Lee Building
239 North Lamar Street
Jackson, MS 39201
(601) 359-1288

Missouri
Department of Mental Health
Division of Alcoholism and
Drug Abuse
1706 East Elm Street
P.O. Box 687
Jefferson City, MO 65102
(314) 751-4942

Montana
Department of Institutions
Alcohol and Drug Abuse
Division
1539 11th Avenue
Helena, MT 59620
(406) 444-2827

Nebraska
Department of Public
Institutions
Division of Alcoholism and
Drug Abuse
801 West Van Dorn Street
P.O. Box 94728
Lincoln, NB 68509-4728
(402) 471-2851, Ext. 5583

Nevada
Department of Human
Resources
Bureau of Alcohol and Drug
Abuse
505 East King Street
Room 500
Carson City, NV 89710
(702) 687-4790

New Hampshire
Department of Health and
Human Services
Office of Alcohol and Drug
Abuse Prevention
State Office
Park South
105 Pleasant Street
Concord, NH 03301
(603) 271-6100

New Jersey
Department of Health
Division of Alcoholism and
Drug Abuse
129 East Hanover Street CN
362
Trenton, NJ 08625
(609) 292-8949

New Mexico
Health and Environment
Department
Behavioral Health Services
Division/
Substance Abuse
Harold Runnels Building
1190 Saint Francis Drive
Santa Fe, NM 87503
(505) 827-2601

New York
Division of Alcoholism and
Alcohol Abuse
194 Washington Avenue

Albany, NY 12210
(518) 474-5417

Division of Substance Abuse
  Services
Executive Park South
Box 8200
Albany, NY 12203
(518) 457-7629

North Carolina
Department of Human
  Resources
Division of Mental Health,
  Developmental Disabilities,
  and Substance Abuse
  Services
Alcohol and Drug Abuse
  Services
325 North Salisbury Street
Albemarle Building
Raleigh, NC 27603
(919) 733-4670

North Dakota
Department of Human Services
Division of Alcohol and Drug
  Abuse
1839 East Capital Avenue
Bismarck, ND 58501-2152
(701) 224-2769

Ohio
Division of Alcohol and Drug
  Addiction Services
246 North High Street
3rd Floor
Columbus, OH 43266-0170
(614) 466-3445

Oklahoma
Department of Mental Health
  and Substance Abuse
  Services
Alcohol and Drug Abuse
  Services
1200 North East 13th Street
P.O. Box 53277
Oklahoma City, OK 73152-
  3277
(405) 271-8653

Oregon
Department of Human
  Resources

Office of Alcohol and Drug
  Abuse Programs
1178 Chemeketa NE
#102
Salem, OR 97310
(503) 378-2163

Pennsylvania
Department of Health
Office of Drug and Alcohol
  Programs
Health and Welfare Building
Room 809
P.O. Box 90
Harrisburg, PA 17108
(717) 787-9857

Rhode Island
Department of Mental Health,
  Mental Retardation and
  Hospitals
Division of Substance Abuse
Substance Abuse
  Administration Building
P.O. Box 20363
Cranston, RI 02920
(401) 464-2091

South Carolina
Commission on Alcohol and
  Drug Abuse
3700 Forest Drive
Suite 300
Columbia, SC 29204
(803) 734-9520

South Dakota
Department of Human
  Services
700 Governor's Drive
Pier South D
Pierre, SD 57501-2291
(605) 773-4806

Tennessee
Department of Mental Health
  and Mental Retardation
Alcohol and Drug Abuse
  Services
706 Church Street
Nashville, TN 37243-0675
(615) 741-1921

Texas
Commission on Alcohol and
  Drug Abuse

720 Bracos Street
Suite 403
Austin, TX 78701
(512) 463-5510

Utah
Department of Social Services
Division of Substance Abuse
120 North 200 West
4th Floor
Salt Lake City, UT 84103
(801) 538-3939

Vermont
Agency of Human Services
Department of Social and
  Rehabilitation Services
Office of Alcohol and Drug
  Abuse Programs
103 South Main Street
Waterbury, VT 05676
(802) 241-2170

Virginia
Department of Mental Health
  and Mental Retardation
Division of Substance Abuse
109 Governor Street
8th Floor
P.O. Box 1797
Richmond, VA 23214
(804) 786-5313

Washington
Department of Social and
  Health Service
Division of Alcohol and
  Substance Abuse
12th and Franklin
Mail Stop OB 21W
Olympia, WA 98504
(206) 753-5866

West Virginia
Department of Health and
  Human Resources
Office of Behavioral Health
  Services
Division on Alcoholism and
  Drug Abuse
Capital Complex
1900 Kanawha Boulevard East
Building 3, Room 402
Charleston, WV 25305
(304) 348-2276

Wisconsin
Department of Health and
Social Services
Division of Community
Services
Bureau of Community
Programs
Office of Alcohol and Drug
Abuse
1 West Wilson Street
P.O. Box 7851
Madison, WI 53707-7851
(608) 266-2717

Wyoming
Alcohol And Drug Abuse
Programs
451 Hathaway Building
Cheyenne, WY 82002
(307) 777-7115

U.S. TERRITORIES AND POSSESSIONS

American Samoa
LBJ Tropical Medical Center
Department of Mental Health
Clinic
Pago Pago, American Samoa
96799

Guam
Mental Health & Substance
Abuse Agency
P.O. Box 20999
Guam 96921

Puerto Rico
Department of Addiction
Control Services
Alcohol and Drug Abuse
Programs
Avenida Barbosa
P.O. Box 414
Rio Piedras, PR 00928-1474
(809) 763-7575

Trust Territories
Director of Health Services
Office of the High
Commissioner
Saipan, Trust Territories
96950

Virgin Islands
Division of Health and
Substance Abuse
Becastro Building
3rd Street, Sugar Estate
St. Thomas, Virgin Islands
00802

CANADA
Canadian Centre on
Substance Abuse
112 Kent Street, Suite 480
Ottawa, Ontario
K1P 5P2
(613) 235-4048

Alberta
Alberta Alcohol and Drug
Abuse Commission
10909 Jasper Avenue, 6th
Floor
Edmonton, Alberta
T5J 3M9
(403) 427-2837

British Columbia
Ministry of Labour and
Consumer Services
Alcohol and Drug Programs
1019 Wharf Street, 5th Floor
Victoria, British Columbia
V8V 1X4
(604) 387-5870

Manitoba
The Alcoholism Foundation of
Manitoba
1031 Portage Avenue
Winnipeg, Manitoba
R3G 0R8
(204) 944-6226

New Brunswick
Alcoholism and Drug
Dependency Commission
of New Brunswick
65 Brunswick Street
P.O. Box 6000
Fredericton, New Brunswick
E3B 5H1
(506) 453-2136

Newfoundland
The Alcohol and Drug
Dependency Commission
of Newfoundland and
Labrador
Suite 105, Prince Charles
Building
120 Torbay Road, 1st Floor
St. John's, Newfoundland
A1A 2G8
(709) 737-3600

Northwest Territories
Alcohol and Drug Services
Department of Social Services
Government of Northwest
Territories
Box 1320 - 52nd Street
6th Floor, Precambrian
Building
Yellowknife, Northwest
Territories
S1A 2L9
(403) 920-8005

Nova Scotia
Nova Scotia Commission on
Drug Dependency
6th Floor, Lord Nelson
Building
5675 Spring Garden Road
Halifax, Nova Scotia
B3J 1H1
(902) 424-4270

Ontario
Addiction Research
Foundation
33 Russell Street
Toronto, Ontario
M5S 2S1
(416) 595-6000

Prince Edward Island
Addiction Services of Prince
Edward Island
P.O. Box 37
Eric Found Building
65 McGill Avenue
Charlottetown, Prince Edward
Island
C1A 7K2
(902) 368-4120

Quebec
Service des Programmes aux
   Personnes Toxicomanie
Gouvernement du Quebec
Ministere de la Sante et des
   Services Sociaux
1005 Chemin Ste. Foy
Quebec City, Quebec
G1S 4N4
(418) 643-9887

Saskatchewan
Saskatchewan Alcohol and
   Drug Abuse Commission
1942 Hamilton Street
Regina, Saskatchewan
S4P 3V7
(306) 787-4085

Yukon
Alcohol and Drug Services
Department of Health and
   Social Resources
Yukon Territorial Government
6118-6th Avenue
P.O. Box 2703
Whitehorse, Yukon Territory
Y1A 2C6
(403) 667-5777

Acknowledgments

FIGURE 2. J. E. Henningfield, M. L. Stitzer, and R. R. Griffiths, *Addictive Behaviors* 5 (1980): 265–72. Reprinted by permission of Pergamon Press, Ltd.

FIGURES 9, 10. J. E. Henningfield and R. R. Griffiths, *Behavior Research Methods and Instrumentation* 11 (1979): 538–44. Reprinted by permission of The Psychonomic Society.

FIGURE 11. J. E. Henningfield and R. R. Griffiths, *Clinical Pharmacology and Therapeutics* 30 (1981): 497–505. Reprinted by permission of the C. V. Mosby Company.

FIGURE 13. J. E. Henningfield. In *Advances in Behavioral Pharmacology*, vol. IV, (eds.) T. Thompson and P. B. Dews. New York: Academic Press, 1984, pp. 131–210. Reprinted by permission.

FIGURE 14. W. A. Hunt, L. W. Barnett, and L. G. Branch, *Journal of Clinical Psychology* 27 (1971): 455–56. Reprinted by permission.

W. A. Hunt and W. R. General, *Journal of Comparative Psychology* 1 (1973): 66–68. Reprinted by permission.

FIGURE 15. D. R. Jasinski, R. E. Johnson, and J. E. Henningfield, *Trends in Pharmacological Sciences* 5 (1984) 196–200. Reprinted by permission of Elsevier Biomedical Press.

FIGURE 16. J. E. Henningfield, K. Miyasato, and D. R. Jasinski, *Pharmacology, Biochemistry, and Behavior* 19 (1983): 887–90. Reprinted by permission of ANKHO International Inc.

Further Reading

General

Berger, Gilda. *Drug Abuse: The Impact on Society.* New York: Watts, 1988. (Gr. 7–12)

Cohen, Susan, and Daniel Cohen. *What You Can Believe About Drugs: An Honest and Unhysterical Guide for Teens.* New York: M. Evans, 1987. (Gr. 7–12)

Musto, David F. *The American Disease: Origins of Narcotic Control.* Rev. ed. New Haven: Yale University Press, 1987.

National Institute on Drug Abuse. *Drug Use, Drinking, and Smoking: National Survey Results from High School, College, and Young Adult Populations, 1975–1988.* Washington, DC: Public Health Service, Department of Health and Human Services, 1989.

O'Brien, Robert, and Sidney Cohen. *Encyclopedia of Drug Abuse.* New York: Facts on File, 1984.

Snyder, Solomon H., M.D. *Drugs and the Brain.* New York: Scientific American Books, 1986.

U.S. Department of Justice. *Drugs of Abuse.* 1989 edition. Washington, DC: Government Printing Office, 1989.

Nicotine

American Journal of Public Health, February 1989. Special issue: "Tobacco and Health."

Amos, A., and C. Chollat-Traquet. "Women and Tobacco." *World Health*, April-May 1990.

Colosi, Marco L. "Do Employees Have the Right to Smoke?" *Personnel Journal*, April 1988.

Gibbs, Nancy R. "All Fired Up Over Smoking." *Time*, April 18, 1988.

Gloeckner, C. "Lung Cancer—the Smoking Gun." *Current Health*, March 1989.

National Research Council. *Environmental Tobacco Smoke: Measuring Exposures and Assessing Health Effects.* Washington, DC: National Academy Press, 1986.

"The Nation's Health Bill for Smoking: $52 Billion." *Public Health Report*, May-June 1990.

Raloff, J. "Pictures Show Smoking's Ill Effects on DNA." *Science News*, March 11, 1989.

Schmeisser, Peter. "Pushing Cigarettes Overseas." *The New York Times Magazine*, July 10, 1988.

Steele, W. "The Downside of Smoking Tobacco and Marijuana." *Current Health*, November 1989.

"Tobacco Industry: On the Defensive, But Still Going Strong." Washington, DC: Editorial Research Reports, September 21, 1990.

U.S. Surgeon General. *The Health Consequences of Smoking: Nicotine Addiction.* Washington, DC: Department of Health and Human Services, 1988.

————. *Reducing the Health Consequences of Smoking: 25 Years of Progress.* Washington, DC: Department of Health and Human Services, 1989.

Warner, Kenneth E. *Selling Smoke: Cigarette Advertising and Public Health.* Washington, DC: American Public Health Association, 1986.

"Who Smokes, Who Starts—and Why." Washington, DC: Editorial Research Reports, March 24, 1989.

Glossary

addiction a condition caused by repeated drug use and characterized by a compulsive urge to continue using the drug, a tendency to increase the dosage, and physiological and/or psychological dependence

AIDS acquired immune deficiency syndrome; an acquired defect in the immune system; the final stage of the disease caused by the human immunodeficiency virus (HIV); spread by the exchange of blood (including on contaminated hypodermic needles), by sexual contact, through nutritive fluids passed from a mother to her fetus, and through breast-feeding; leaves victims vulnerable to certain, often fatal, infections and cancers

alveoli tiny air holes in the lungs that fill up with tobacco by-products (tar, for example) during smoking, causing respiratory diseases

carcinogenic causing cancer

cardiovascular of, related to, or involving the heart and blood vessels

chronic obstructive pulmonary diseases (COPDs) several diseases of the respiratory system (primarily emphysema and bronchitis) that are usually caused by smoking and collectively are the fifth leading cause of death in the United States

euphoria a feeling of well-being or elation

National Institute on Drug Abuse (NIDA) a subdivision of the National Institutes of Health that monitors patterns of drug use and research in the area of drug abuse

Nicorette a chewing gum containing nicotine that is used to help people stop smoking

nicotine a drug occurring naturally in tobacco leaves and having both stimulating and depressing effects on the body; it is the addictive ingredient in tobacco

smokeless tobacco products made from tobacco that can be used without burning; these include snuff, chewing tobacco, and a synthetic cigarette that can be "puffed" without lighting

tar a thick black material that looks like road tar and is produced when tobacco is burned; tar has been shown to cause cancer in laboratory animals

tolerance a decrease of susceptibility to the effects of a drug due to its continued administration, resulting in the user's need to increase the drug dosage in order to achieve the effects experienced previously.

Index

Jack E. Henningfield, Ph.D., is chief of the Clinical Pharmacology Branch of the Addiction Research Center at the National Institute on Drug Abuse. He is also an associate professor of behavioral biology in the Department of Psychiatry and Behavioral Sciences at the Johns Hopkins University School of Medicine in Baltimore and is an adjunct professor in the toxicology program at the University of Maryland at Baltimore. Dr. Henningfield was a scientific editor of the 1988 *Surgeon General's Report (Nicotine Addiction)* and was a scientific consultant on the Surgeon General's report on smokeless tobacco (1986).

Paul R. Sanberg, Ph.D., is a professor of psychiatry, psychology, neurosurgery, physiology, and biophysics at the University of Cincinnati College of Medicine. Currently he is also a professor of psychiatry at Brown University and scientific director for Cellular Transplants, Inc., in Providence, Rhode Island.

Professor Sanberg has held research positions at the Australian National University at Canberra, the Johns Hopkins University School of Medicine, and Ohio University. He has written many journal articles and book chapters in the fields of neuroscience and psychopharmacology. He has served on the editorial boards of many scientific journals and is the recipient of numerous awards.

Solomon H. Snyder, M.D., is Distinguished Service Professor of Neuroscience, Pharmacology and Psychiatry at the Johns Hopkins University School of Medicine. He has served as president of the Society for Neuroscience and in 1978 received the Albert Lasker Award in Medical Research. He has authored *Drugs and the Brain, Uses of Marijuana, Madness and the Brain, The Troubled Mind,* and *Biological Aspects of Mental Disorder* and has edited *Perspectives in Neuropharmacology: A Tribute to Julius Axelrod.* Professor Snyder was a research associate with Dr. Axelrod at the National Institutes of Health.

Barry L. Jacobs, Ph.D., is currently a professor in the neuroscience program at Princeton University. Professor Jacobs is the author of *Serotonin Neurotransmission and Behavior* and *Hallucinogens: Neurochemical, Behavioral and Clinical Perspectives.* He has written many journal articles in the field of neuroscience and contributed numerous chapters to the books on behavior and brain science. He has been a member of several panels of the National Institute of Mental Health.

Jerome H. Jaffe, M.D., formerly professor of psychiatry at the College of Physicians and Surgeons, Columbia University, is director of the Addiction Research Center of the National Institute on Drug Abuse. Dr. Jaffe is also a psychopharmacologist and has conducted research on a wide range of addictive drugs and developed treatment programs for addicts. He has acted as special consultant to the president on narcotics and dangerous drugs and was the first director of the White House Special Action Office for Drug Abuse Prevention.

0 0 6 3 9 9 5

NICOTINE AND OLD FAS
HIONED ADDICT
HENNINGFIELD